MW01297192
SUMMIT MATH
Learn at your OWN pace.
ALGEBRA 2
second edition
6
LINEAR & NONLINEAR SYSTEMS OF EQUATIONS

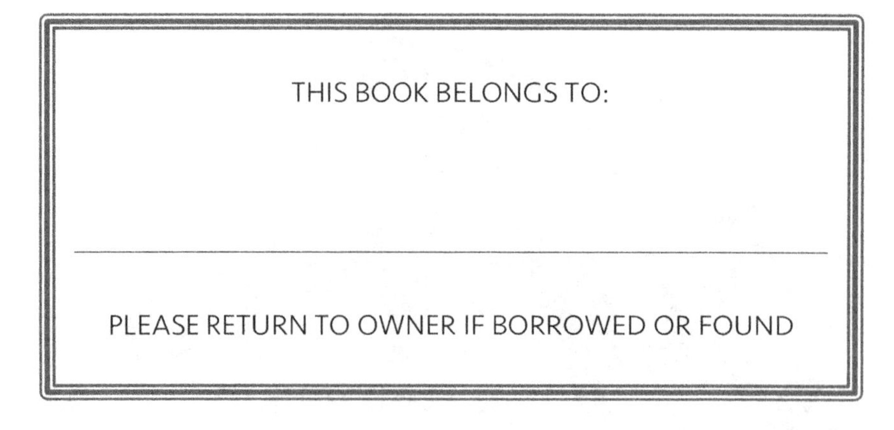

DEDICATION

To Lauren, Chloe, Dawson and Teagan

ACKNOWLEDGEMENTS

I started writing these books in 2013 to help my students learn better. I kept writing them because I received encouraging feedback from students, parents and teachers. Thank you to all who have used these books, pointed out my mistakes, and made suggestions along the way. Thank you to all of the students and parents who asked me to keep writing more books. Thank you to my family for supporting me through every step of this journey.

This book was typeset in the following fonts:
Seravek + Mohave + *Heading Pro*

Graphics in Summit Math books are made using the following resources:
Microsoft Excel | Microsoft Word | Desmos | Geogebra | Adobe Illustrator

First printed in 2017

Printed in the U.S.A.

Summit Math Books are written by Alex Joujan.

www.summitmathbooks.com

Learning math through Guided Discovery:
A Guided Discovery learning experience is designed to help you experience a feeling of discovery as you learn each new topic.

Why this curriculum series is named Summit Math:
Learning through Guided Discovery can be compared to climbing a mountain. Climbing and learning both require effort and persistence. In both activities, people naturally move at different paces, but they can reach the summit if they keep moving forward. Whether you race rapidly through these books or step slowly through each scenario, this curriculum is designed to keep advancing your learning until you reach the end of the book.

Guided Discovery Scenarios:
The Guided Discovery Scenarios in this book are written and arranged to show you that new math concepts are related to previous concepts you have already learned. Try to fully understand each scenario before moving on to the next one. To do this, try the scenario on your own first, check your answer when you finish, and then fix any mistakes, if needed. Making mistakes and struggling are essential parts of the learning process.

Homework and Extra Practice Scenarios:
After you complete the scenarios in each Guided Discovery section, you may think you know those topics well, but over time, you will forget what you have learned. Extra practice will help you develop better retention of each topic. Use the Homework and Extra Practice Scenarios to improve your understanding and to increase your ability to retain what you have learned.

The Answer Key:
The Answer Key is included to promote learning. When you finish a scenario, you can get immediate feedback. When the Answer Key is not enough to help you fully understand a scenario, you should try to get additional guidance from another student or a teacher.

Star symbols:
Scenarios marked with a star symbol ★ can be used to provide you with additional challenges. Star scenarios are like detours on a hiking trail. They take more time, but you may enjoy the experience. If you skip scenarios marked with a star, you will still learn the core concepts of the book.

To learn more about Summit Math and to see more resources:
Visit www.summitmathbooks.com.

GUIDED DISCOVERY SCENARIOS

As you complete scenarios in this part of the book, follow the steps below.

Step 1: Try the scenario.
Read through the scenario on your own or with other classmates. Examine the information carefully. Try to use what you already know to complete the scenario. Be willing to struggle.

Step 2: Check the Answer Key.
When you look at the Answer Key, it will help you see if you fully understand the math concepts involved in that scenario. It may teach you something new. It may show you that you need guidance from someone else.

Step 3: Fix your mistakes, if needed.
If there is something in the scenario that you do not fully understand, do something to help you understand it better. Go back through your work and try to find and fix your errors. Mistakes provide an opportunity to learn. If you need extra guidance, get help from another student or a teacher.

After Step 3, go to the next scenario and repeat this 3-step cycle.

NEED EXTRA HELP? watch videos online

Teaching videos for every scenario in the Guided Discovery section of this book are available at www.summitmathbooks.com/algebra-2-videos.

CONTENTS

Section 1

REVIEW GRAPHING SYSTEMS, SUBSTITUTION AND ELIMINATION

You were introduced to the topic of Systems of Equations in a previous lesson. As you prepare to discover more about this topic, it will help to first review what you have you already learned.

1. Without graphing, where do the two lines intersect? After you use an algebraic method to determine the intersection point, graph the lines to confirm that your intersection point is accurate.

   Line 1: $y = -x + 5$
   Line 2: $y = 2x - 4$

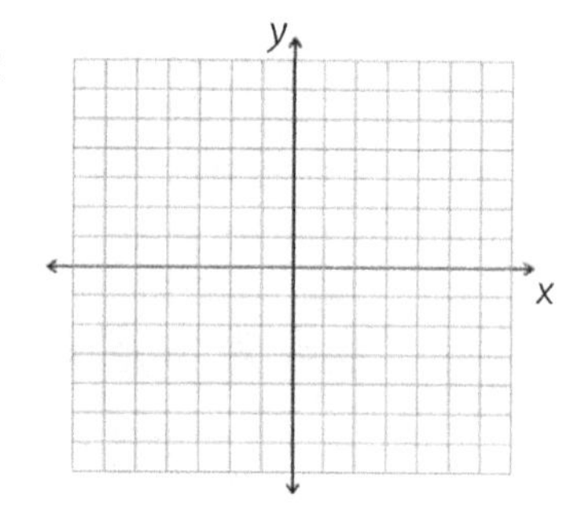

2. Do you remember how to solve a system of equations using the Substitution Method?

   a. In your own words, write out how to use this method.

   b. Use the Substitution Method to show that the two lines given by the equations $2x - 2y = 14$ and $3x + y = 5$ intersect at the point (3, −4).

3. Solve each system using the Substitution Method. As a reminder, you can make a substitution for either *x* or *y*. The goal is to change a 2-variable equation to make it contain only one variable.

   a. $y = 2x + 6$
      $-3x + 6y = -9$

   b. $x = 3y - 12$
      $3y - 2x = 9$

4. Where do two lines intersect if the equation for Line 1 is $y = 2x + 6$ and the equation for Line 2 is $-3x + 6y = -9$?

5. Do you remember how to solve a system of equations using the Elimination Method?

a. In your own words, write out how to use this method.

b. Use the Elimination Method to show that the two lines given by the equations $2x-2y=14$ and $3x+y=5$ intersect at the point (3, -4).

6. Solve each system of equations using the Elimination Method. As a reminder, the equations can either be added or subtracted in order to eliminate one of the variables.

a. $6y-3x=-9$
$y=2x+6$

b. $5x+2y=10$
$6y-2x=-4$

7. Where do two lines intersect if the equation for Line 1 is $5x+2y=10$ and the equation for Line 2 is $6y-2x=-4$?

8. Write the equation of the line that passes through the points (-3, 3) and (6, 0). Graph the line to confirm that it passes through the given points.

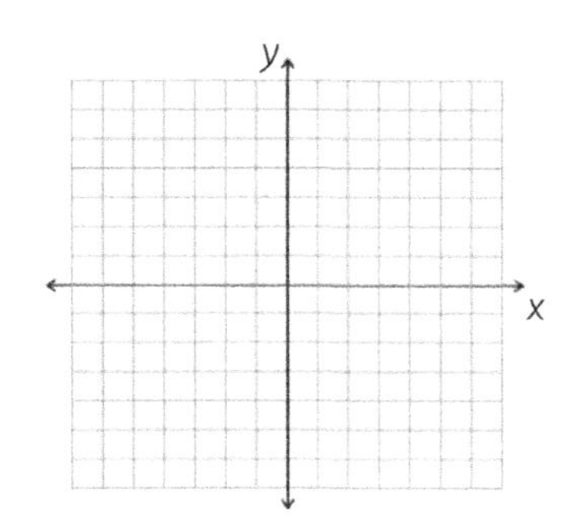

9. Two lines are shown in the graph. Write the equation of one line in Slope-Intercept Form and the other equation in Standard Form.

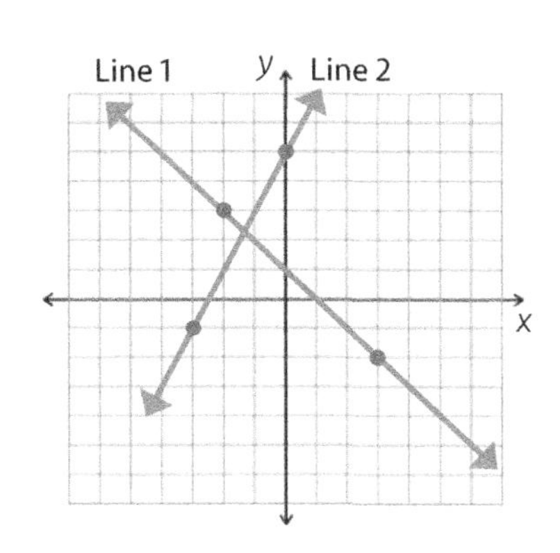

10. Find the exact intersection point of the two lines in the previous scenario.

11. What information do you need to find the equation of a line in Slope-Intercept Form?

12. ★Estimate the coordinates of the intersection point of the two lines and then determine the exact coordinates.

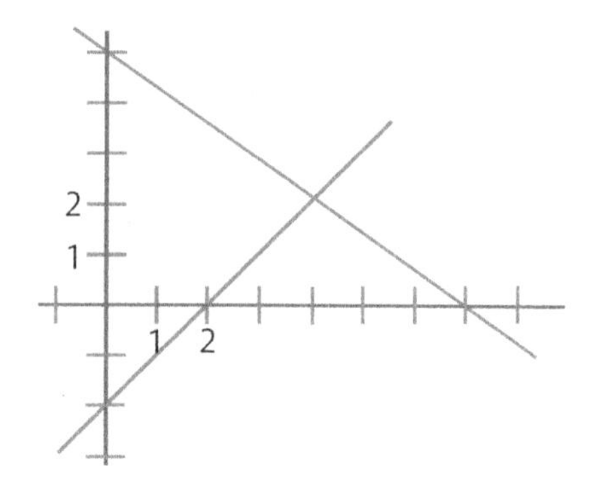

13. If you look at the two lines in the graph above and the x- and y-axes, you will notice that there are triangles formed by these converging lines and axes.

a. What is the height of the triangle that is formed by the x-axis and the two lines?

b. What is the height of the triangle that is bounded by the y-axis and the two lines?

14. Draw a second line on each graph shown below to create a system of linear equations that has the stated solution.

a. one solution at (3, 2)

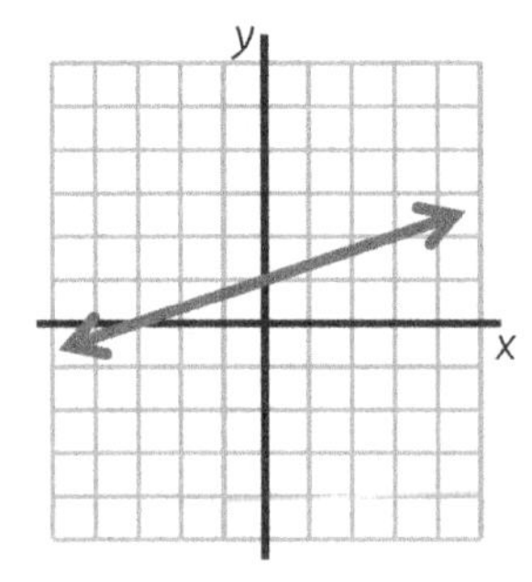

b. no solution

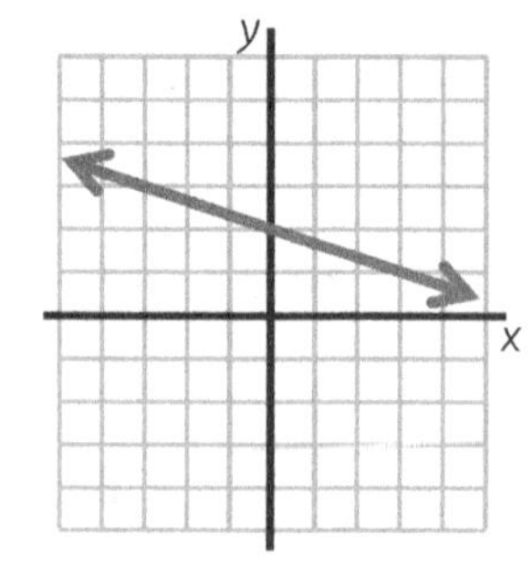

15. For each of the graphs in the previous scenario, identify the equation of the original line and the line that you drew. Write the equations in Slope-Intercept Form.

16. Without graphing them, where do the two lines intersect?

Line 1: $6 + 4x = -2y$

Line 2: $2x + y = 6$

17. The previous two lines do not intersect. Why is this?

18. Without graphing them, where do the two lines intersect?

Line 1: $y = \frac{3}{2}x + 4$

Line 2: $6x - 4y = -16$

19. Consider the line shown in the graph. Draw a second line that is perpendicular to the original one and passes through the point (−6, 5).

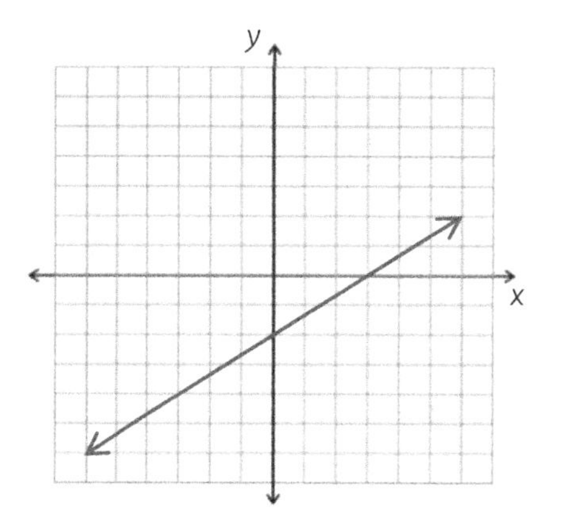

a. How can you determine that two lines are perpendicular by looking at their graphs?

b. Identify the equations of both lines.

20. ★Start with the line formed by the equation $2y - 3x = 18$. Identify a second equation that forms perpendicular lines with identical x-intercepts. Graph both of these lines to confirm that your second equation satisfies the stated condition.

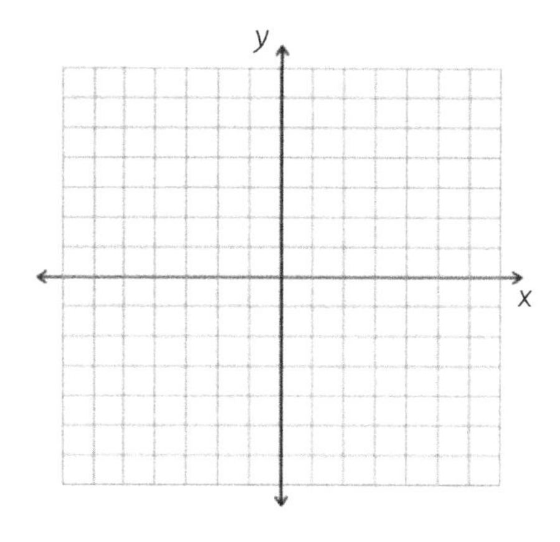

21. Do the three lines $5x - y = 7$, $x + 3y = 11$, and $-3x + 4y = 6$ have a common point of intersection? If so, find it. If not, explain why not.

22. Graph the previous three lines in such a way that their point of intersection is visible on the graph.

23. ★Given the system of equations below, determine the value of the product xy.

$$3x + 2y = 4x - 16$$
$$4y = 8 - 2x$$

24. Calculate the area of the region contained between the x-axis and the graphs of the equations $y - 3x = 0$ and $y + x = 8$.

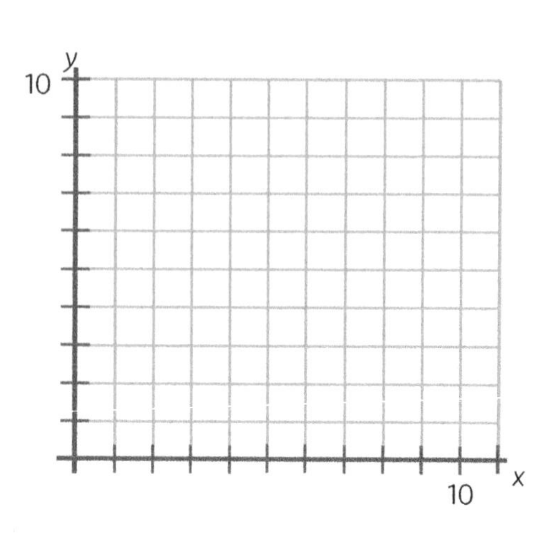

25. Calculate the area of the region that is contained between the y-axis and the graphs of the lines in the previous scenario.

26. Calculate the area of each figure shown below.

a.

4 cm

10 cm

b.

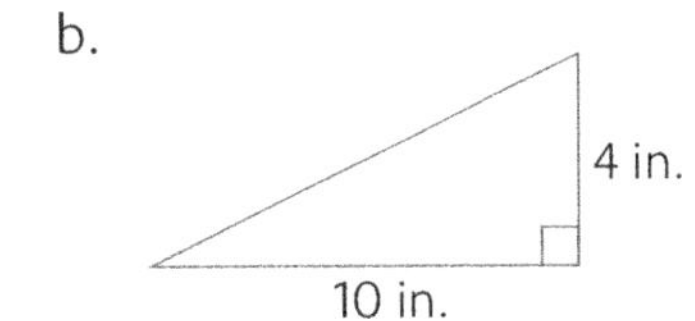

c.

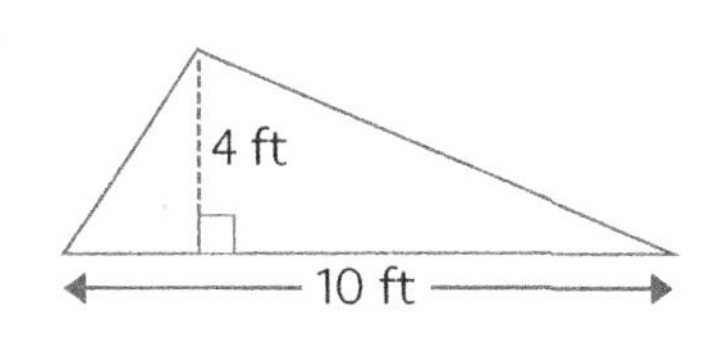

27. The figure shows the graphs of two lines, whose axis intercepts are integers. There is a third line drawn along the x-axis that forms a triangle with these other two lines.

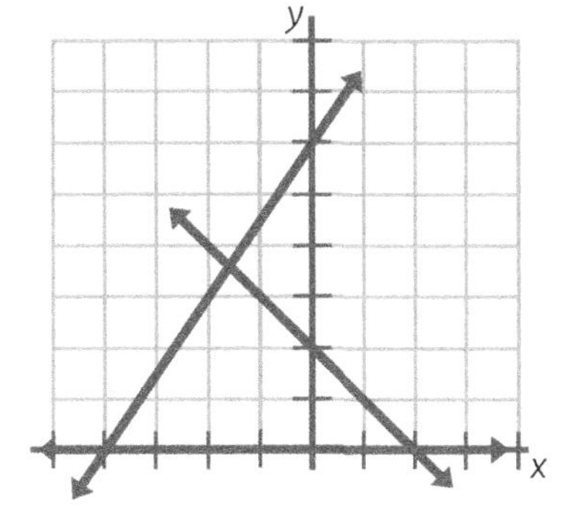

a. Estimate the area of the triangle by counting the squares inside the triangle.

b. Calculate the exact area of the triangle enclosed by the three lines.

28. ★Consider the triangle formed by the three intersecting lines. Assume that points that look like lattice points are lattice points. Coordinates of a lattice point are integers.

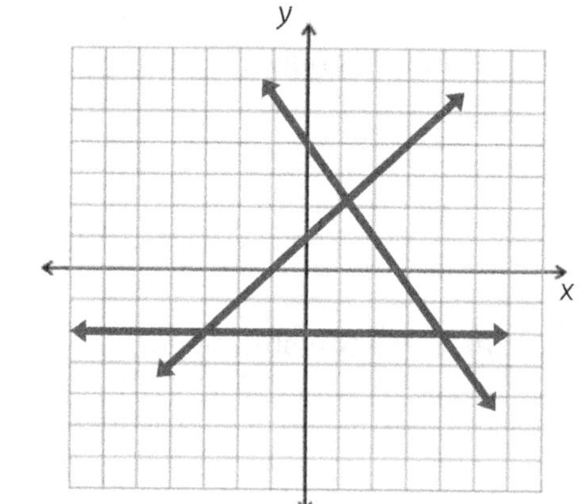

a. Estimate the area of the triangle by counting the squares.

b. Calculate the exact area of the triangle enclosed by the three lines.

29. ★Line H passes through the points (−2, −2) and (0, 4). Line K is perpendicular to Line H and it passes through the point (3, 5). Where do these two lines intersect?

Use this page to record important ideas in the previous section or for any other writing that helps you learn the topics in this book.

Section 2
SCENARIOS INVOLVING LINEAR SYSTEMS

30. A researcher in London is measuring daily temperature changes worldwide. He notices that as the temperature rises in the morning in Washington, D.C., it is falling in the evening in Beijing. For six hours, both temperature changes follow a linear pattern. In Beijing, this linear pattern is modeled by the equation $T=-2h+45$, while the pattern in Washington, D.C. is represented by the equation $T=h+30$. In both equations, T represents the temperature, in Fahrenheit, h hours after the researcher started making his observations.

    a. What are the temperatures in Washington, D.C. and Beijing, respectively, at the moment the researcher starts making his observations?

    b. After how many hours does the researcher observe that the temperature in Beijing is identical to the temperature in Washington D.C.?

31. A family is considering two vacation options. The first option has $1,200 in travel costs and would cost $125 per day for other expenses like housing and meals. The second option has $1,500 in travel costs and would cost $95 per day for other expenses.

    a. Which option is less expensive if you want a 7-day vacation?

    b. You calculate the total expenses on your own and find out that both options cost the same amount. How long is the vacation that you are planning?

32. Haley's grandmother loves to buy dinner for a large group of Haley's friends every once in awhile. She is very generous and buys whatever she thinks they will eat. On one occasion, she buys 6 cheese pizzas and 12 regular hamburgers and pays $133.50. On another occasion, she buys 8 cheese pizzas and 7 regular hamburgers and pays $148.75. What price is she paying for each cheese pizza and each regular hamburger?

33. The two equations show the price (in dollars) of a cup of frozen yogurt from Menchie's and Yogoberry, respectively, if you buy n combined ounces of frozen yogurt and toppings.

$$M = 0.25n + 0.90$$
$$Y = 0.3n + 0.35$$

a. How heavy would a bowl of yogurt need to be in order for it to cost the same amount at both Menchie's and Yogoberry?

b. Which company charges more per ounce of yogurt?

34. At the grocery store, raisins cost $4.25 per pound and almonds cost $7.50 per pound. When you order a 5-pound mixture that combines the two together in one bag, the store charges you $25.15. How many pounds of raisins and how many pounds of almonds were combined to create this mixture?

35. For several years, you have been setting aside nickels and dimes and placing them in a jar. Once the jar fills up, you deposit the coins in an automatic coin-counting machine. The machine prints a receipt stating that you have deposited 580 coins for a total value of $41.80.

a. How many coins of each type did you deposit?

b. If the machine keeps 5% of your money as payment for allowing you to use the quick-counting money machine, how much do you take home with you?

36. ★For your birthday, your aunt, uncle, and grandfather all combined their money to open a savings account for you. They opened your account with an initial balance of $540. If your grandfather's portion was 25% more than the combined value of the portion from your aunt and the portion from your uncle, how much money did your grandfather contribute to your savings account?

37. ★Maria, Rose, and Lucy have had the same sales job for 1 year, earning a flat annual salary of $30,000. Their boss feels that they each deserve a raise so he calls them in one at a time and offers two salary options. Option Alpha is an annual salary of $11,500 with an additional commission of 8% of total annual sales. Option Beta is a $25,000 annual salary with an additional commission of 4% of total annual sales.

Name	Projected Annual Sales Next Year	Highest Possible Salary Next Year
Maria	$402,000	
Rose	$321,500	
Lucy	$____,500	

a. Which salary option will Maria choose? Which salary option will Rose choose?

b. Lucy says it is impossible for her to choose one option over the other. What are her projected annual sales next year?

c. Fill in the chart completely.

38. Two different candles are lit. The graph shows how the height of each candle changes as it burns.

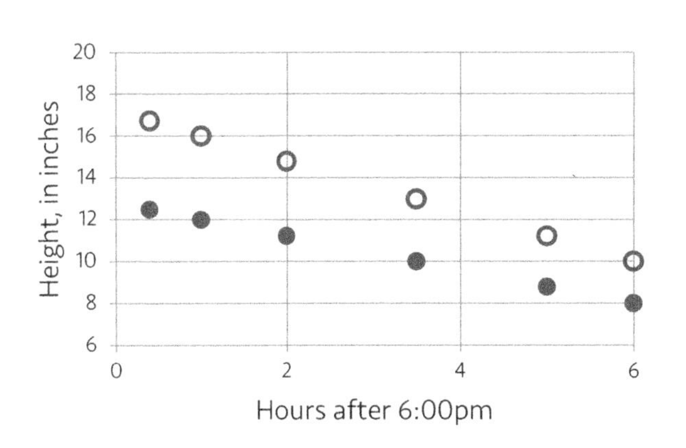

a. How many hours after 6:00pm will the two candles have the same height?

b. At what time will the candles have the same height?

39. Between 2007 and 2015, the popularity of mobile computing devices increased at a faster pace than the popularity of desktop computing devices.

Draw estimated trend lines for the mobile user data and the desktop user data.

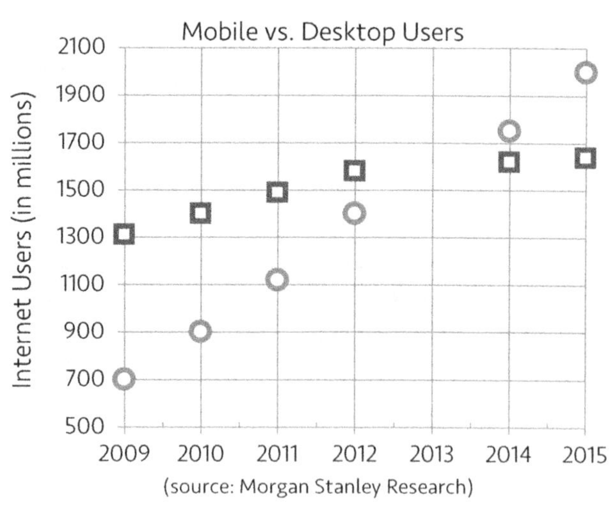

40. ★Estimate the slope of each line in the previous scenario, rounded to the nearest integer. What is the significance of finding the slopes of these lines?

41. The previous scenarios involve systems of linear equations.

a. A linear system cannot have more than one solution because if two lines intersect, they can only intersect at ____ point.

b. Describe the graph of a linear system that has no solution.

c. Describe the graph of a linear system that has infinite solutions.

Use this page to record important ideas in the previous section or
for any other writing that helps you learn the topics in this book.

Section 3
SYSTEMS OF LINEAR INEQUALITIES

In a previous lesson, you learned how to graph a linear inequality. Let's return to this topic.

42. In the Cartesian plane below, plot all of the ordered pairs that have an x-value of 3. How many points can you find?

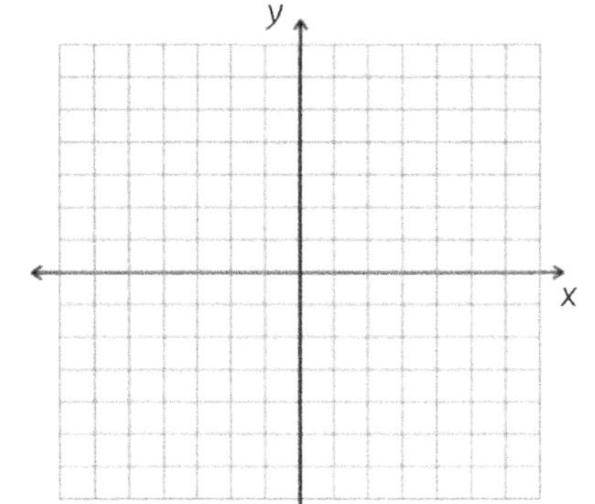

43. In the same plane to the right, graph all of the ordered pairs that have an x-value that is less than 3. How many points can you find?

When you combine the previous two scenarios, they form the solution of the inequality $x \leq 3$. To show this solution on a graph, draw the vertical line for $x = 3$ and darken the region to the left of the line. The darkened region shows the points with x-values less than or equal to 3.

44. In the Cartesian plane shown, plot all of the ordered pairs that have a y-value of -2.

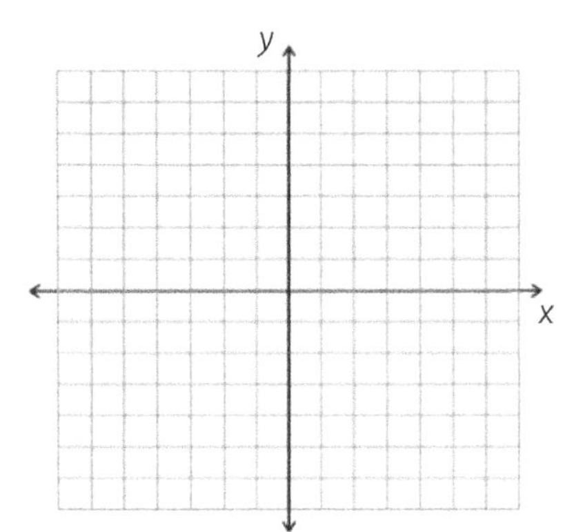

45. In the same plane shown to the right, graph all of the ordered pairs that have a y-value that is greater than -2.

46. When you combine the shaded regions in the previous two scenarios, they form the solution of the inequality $y \geq -2$. How could you change your graph in the previous scenario to display the solution region for the inequality $y > -2$?

47. Graph each inequality.

a. $x > -4$

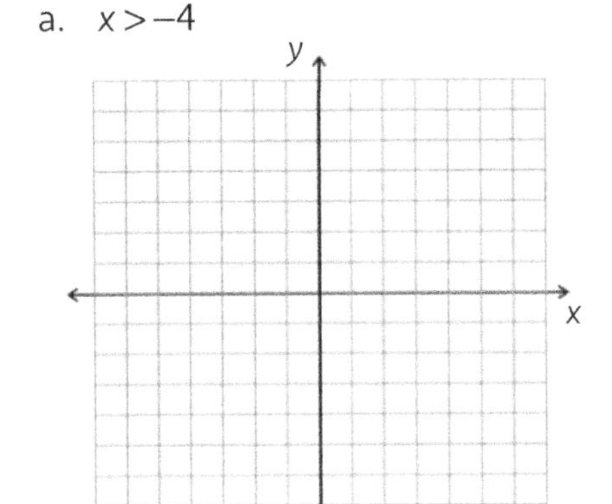

b. $y > 1$

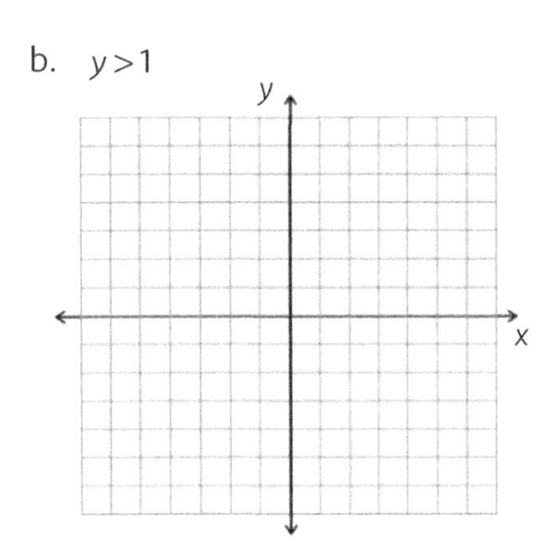

48. Circle the inequalities that would have a shaded region that includes the boundary line.

a. $y<x-3$ b. $y\leq 1-3x$ c. $y+x>6$ d. $2x-5y\leq 10$

49. Graph each inequality.

a. $y<x-3$

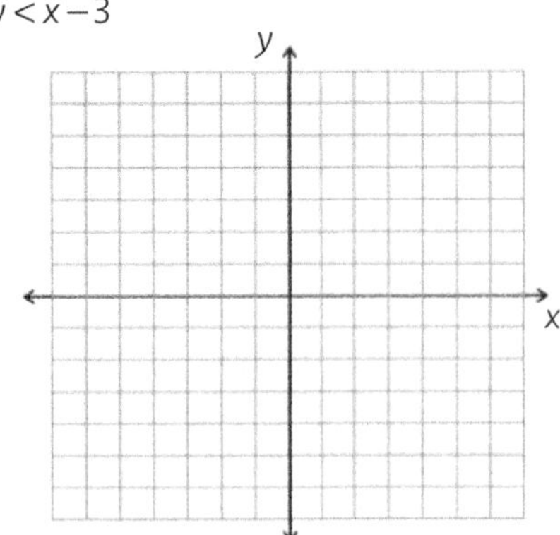

b. $y\geq -2x+3$

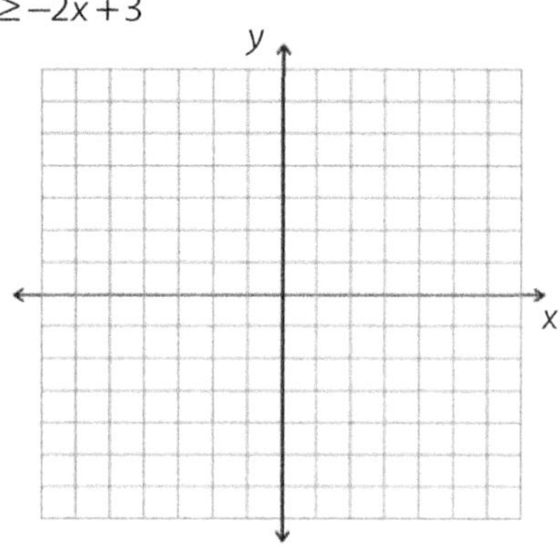

50. Graph the inequality $2x-5y\leq 10$.

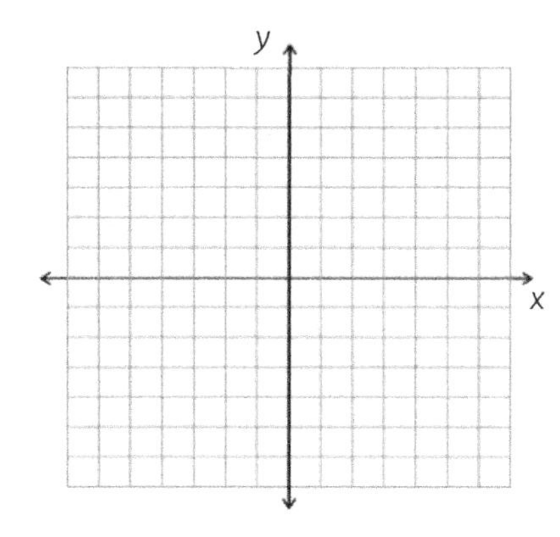

51. Write the inequality that has the solution set shown in each graph below.

a.

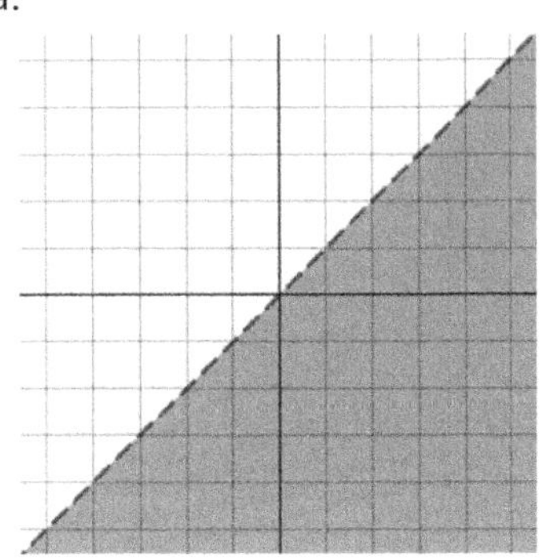

b.

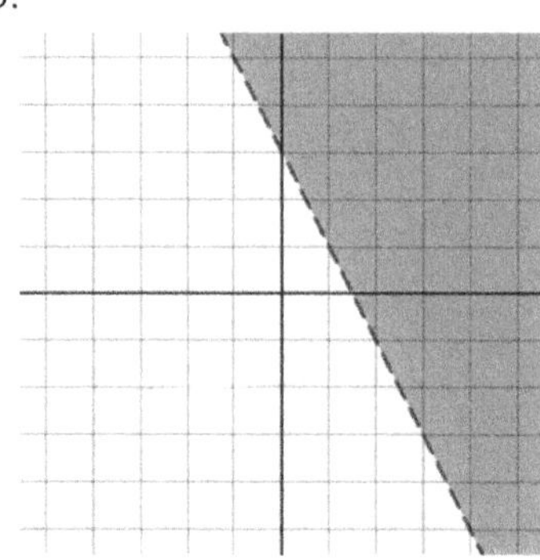

c.

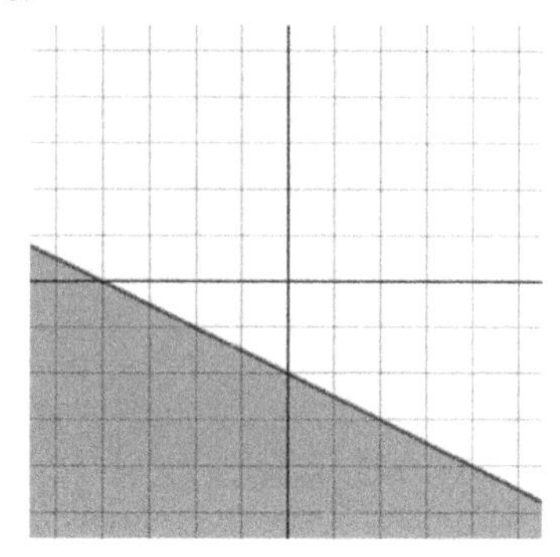

52. Now consider an inequality that is defined by 2 rules.

Rule #1. The x-value of every point is greater than or equal to 1.

Rule #2. The y-value of every point is greater than or equal to 3.

In the graph to the right, plot as many points as you can that satisfy both of the rules.

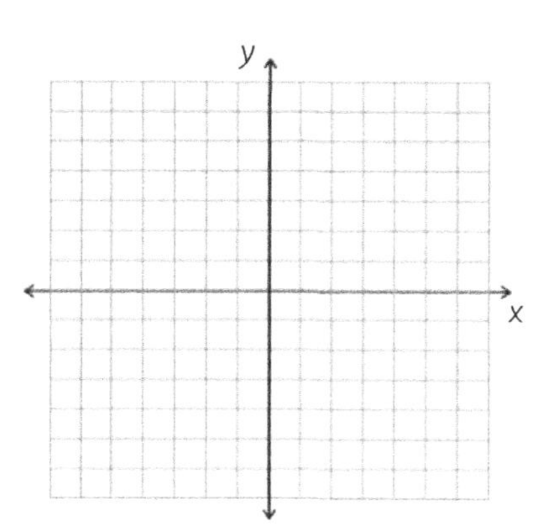

53. The lines in the graph shown divide the plane into 4 regions. Shade in the region that satisfies the 2 inequalities below.

$x \leq -2$

$y \geq -1$

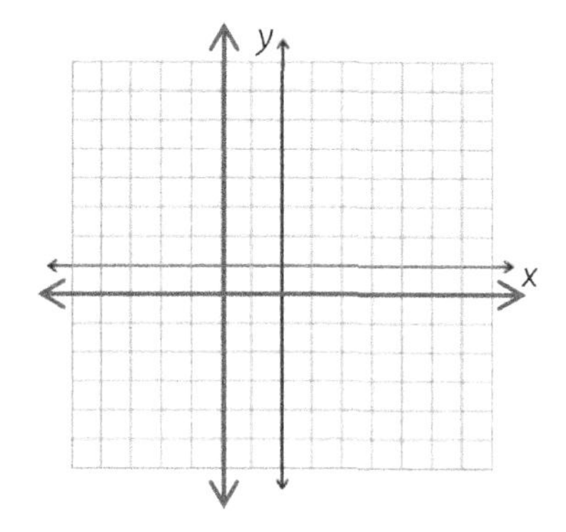

54. The lines in the graph shown divide the plane into 4 regions. Shade in the region that satisfies the 2 inequalities below.

$y \geq x - 3$

$y \geq -x + 2$

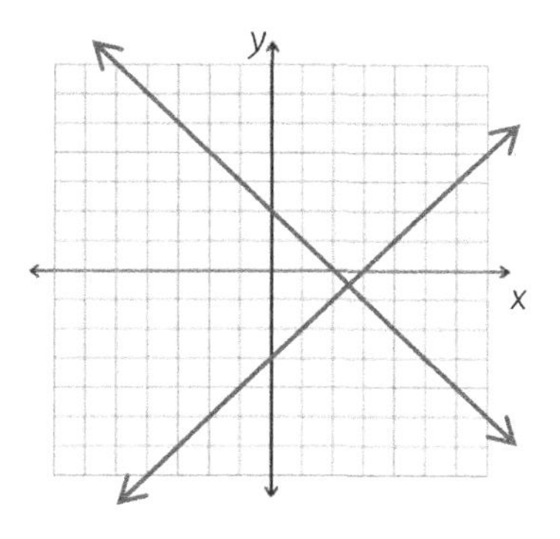

55. How would you change your graph in the previous scenario to display the solution region if the inequalities were changed to become $y > x - 3$ and $y > -x + 2$?

When you graph a linear inequality, the darkened (or shaded) region displays the ordered pairs that make the inequality true. When you graph two linear inequalities, the shaded region displays the ordered pairs that make both inequalities true.

56. Display the solution of the system of inequalities. This is also known as graphing the system.

$y \geq \frac{1}{2}x - 4$

$y < -2x + 1$

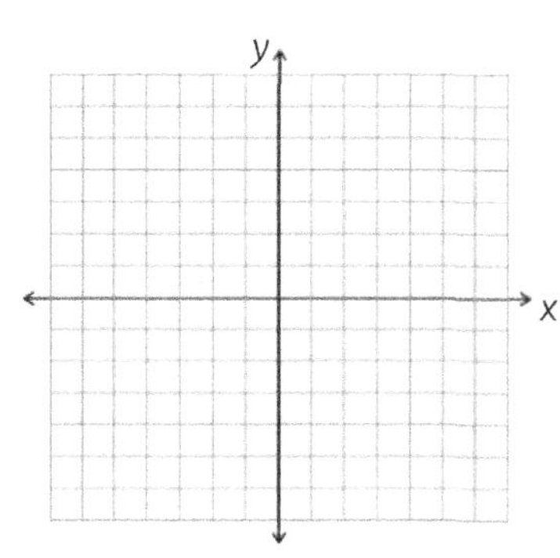

57. Graph the system of inequalities.

$$x+y\leq 4$$
$$x-y<4$$

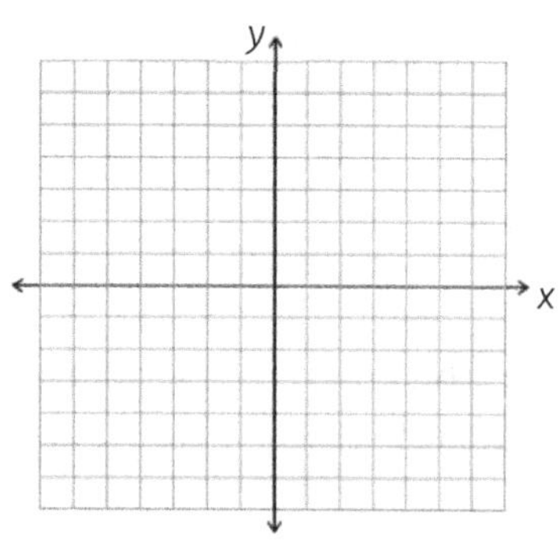

58. Graph the system of inequalities.

$$y<2$$
$$x\geq -3$$
$$x-3y<6$$

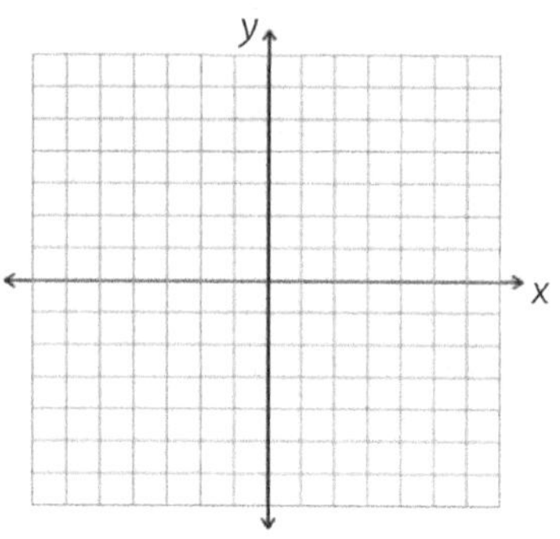

59. Calculate the area of the triangular shaded region in the previous scenario.

60. Write a system of inequalities whose solution set is shown by the shaded region.

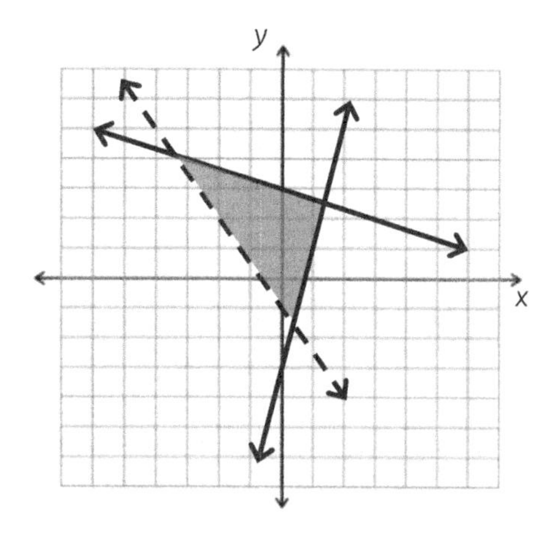

61. The equations of four lines are written below and their graphs are shown to the right.

$3x=5y+20$, $2x+3y=15$, $4x+5y=65$, $3y=2x+15$

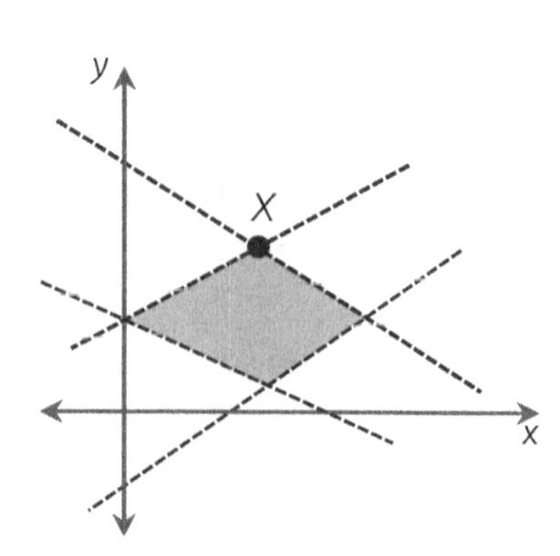

a. Determine the coordinates of X.

b. Write a system of inequalities for the shaded region.

62. Calculate the area of the region defined by the following inequalities.

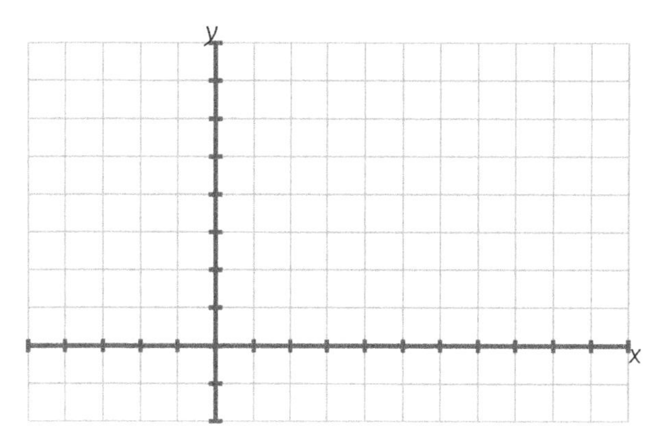

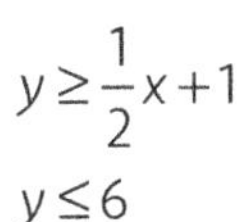

$$y \geq \frac{1}{2}x + 1$$
$$y \leq 6$$
$$x + y > 5$$

63. A grandfather takes his grandchildren to the county fair. He wants to spend at least \$15 but no more than \$30 on snacks. Hot dogs are \$3 each, while popcorn costs \$5 for a bag. How many hot dogs and bags of popcorn can he buy?

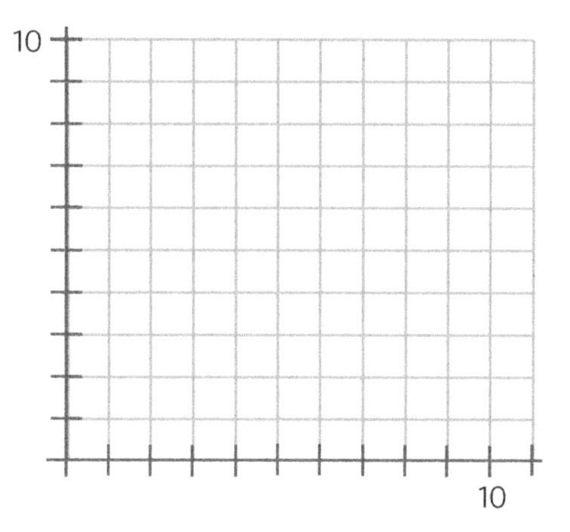

a. Write a system of inequalities that describes this situation.

b. Graph the system to show all possible solutions.

c. Write two possible solutions to the scenario.

d. Where can you look in the shaded region to determine possible solutions?

e. If 1 bag of popcorn and 4 hot dogs is an example of one possible combination, how many different combinations are possible?

64. Create a system of inequalities that matches the description below. Graph the system to check your solution.

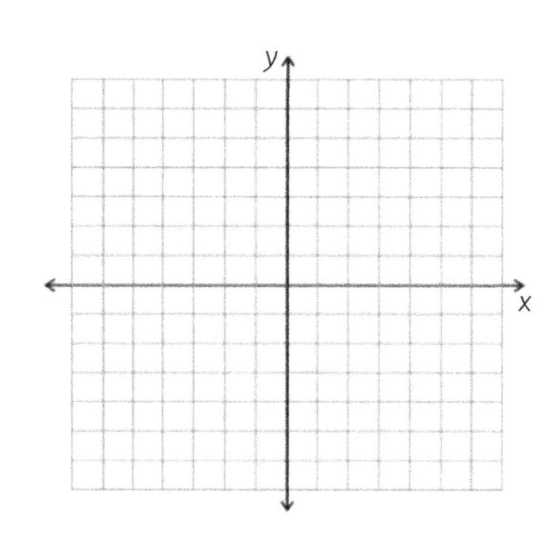

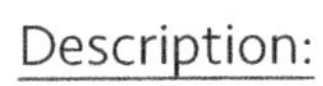

Description:
The shaded region forms a square.
One of the inequalities is $x + 2y < 4$.
One corner of the square is located at $(2, -4)$.

Use this page to record important ideas in the previous section or for any other writing that helps you learn the topics in this book.

Section 4
NONLINEAR SYSTEMS

In the next scenarios, you will see systems that involve nonlinear equations. You can find the solutions to these systems by using methods that you have already learned.

65. Simplify each expression below.

a. $(x+4)^2$

b. $(f-1)^2$

c. $(y^2-10)^2$

66. If $y=-\frac{4}{5}x$, what is the value of y^2?

67. If $x=-3$, what is the value of $-x^2+2x+7$?

68. What does the solution to a system of equations show you about the graphs of those equations?

69. Consider the system shown to the right. How many solutions does this system have? Write the approximate solutions of this system.

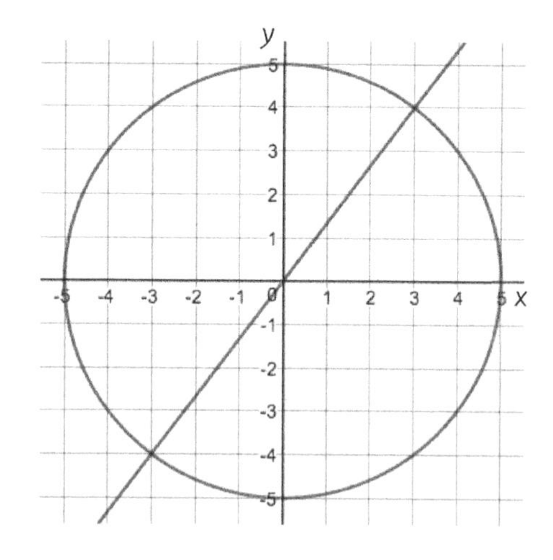

70. Consider the system shown to the right. Identify the <u>approximate</u> solutions of this system.

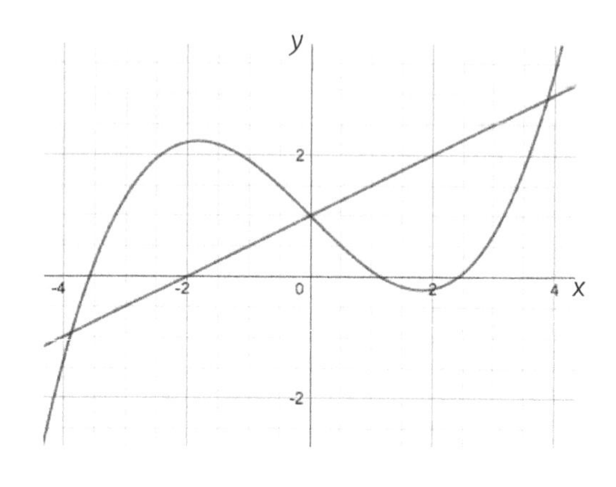

71. The equations for a line and a parabola are shown below. The graph of the line intersects the graph of the parabola at two points, $(-4,\_\_\_)$ and $(3,\_\_\_)$. Find the y-values of the 2 points.

$$y = x - 3$$
$$y = -x^2 + 9$$

72. Two functions are shown below. Where do they intersect?

$$y = -x^2 - x + 5$$
$$y = x^2 - x - 3$$

73. A system of two equations is shown below. A portion of their graphs is also shown in the Cartesian plane to the right. Use algebra to find the exact the solutions of this system. You must verify your result with work.

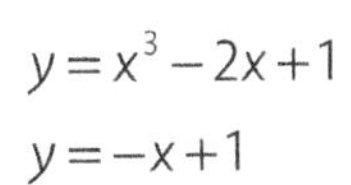

$$y = x^3 - 2x + 1$$
$$y = -x + 1$$

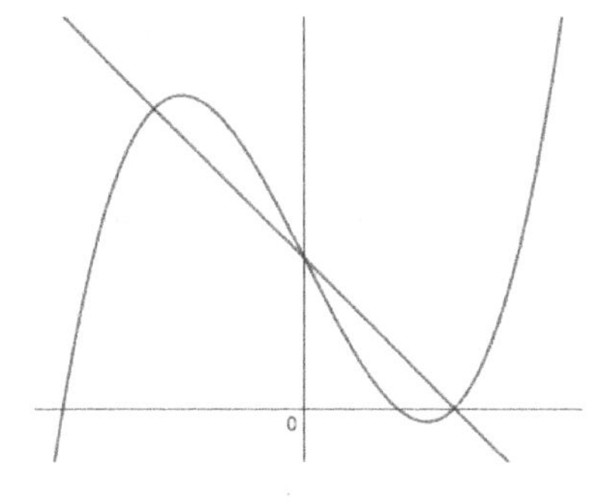

74. Two parabolas are shown in the graph. The equation of the dashed parabola is $y = \frac{1}{4}x^2 - 4$.

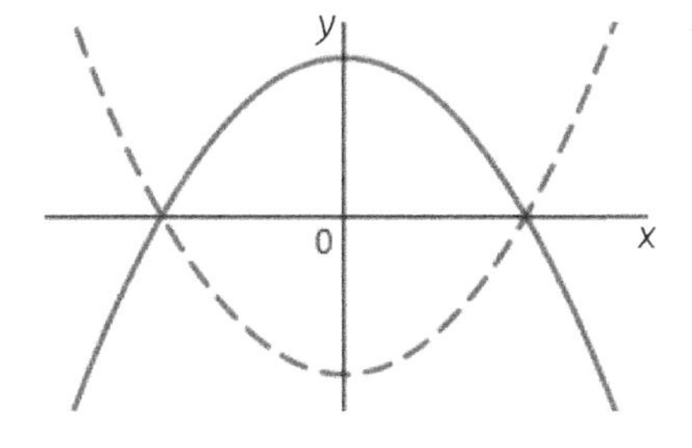

a. What is the equation of the other parabola?

b. If the two parabolas intersect at $(A, 0)$ and $(-A, 0)$, what is the value of A?

75. Where do the two functions intersect? Determine the exact coordinates. Verify the accuracy of your results by displaying the functions on a graphing application.

a. $y = x^2 - 2x - 8$
$y = 2x - 3$

b. $f(x) = x^3 - 2x^2 + x - 1$
$g(x) = 7 + x - 2x^2$

76. Solve the system shown below and verify your results by using a graphing application.

$$\begin{cases} x^2 + y^2 = 25 \\ y = x + 1 \end{cases}$$

77. A system of three equations is shown below. A portion of their graphs is also shown in the Cartesian plane to the right.

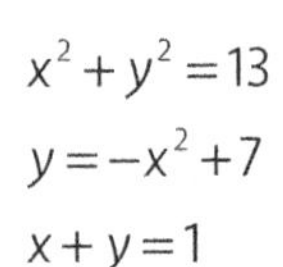

$$x^2 + y^2 = 13$$
$$y = -x^2 + 7$$
$$x + y = 1$$

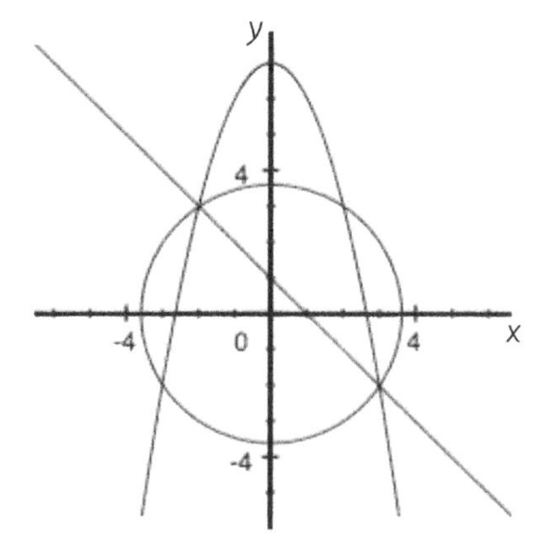

a. How many solutions does this system have?

b. Estimate the solutions of this system.

78. ★Given the system of equations below, what is the value of x^2?

$$\begin{cases} x^2 + y^2 = 117 \\ 2y + 3x = 0 \end{cases}$$

79. Draw a system that consists of a circle and a parabola with the following characteristic.

a. no solution

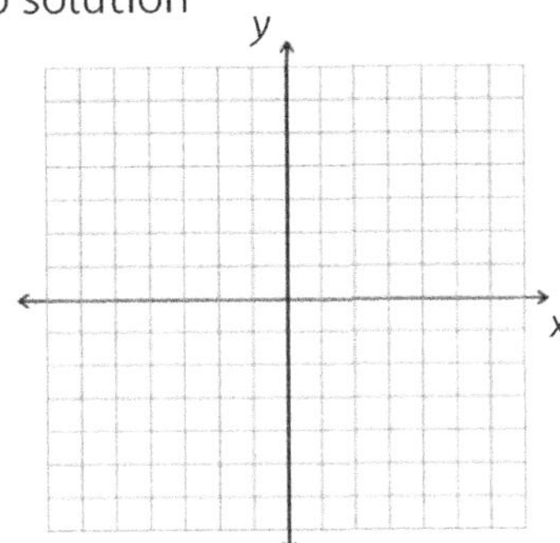

b. one solution

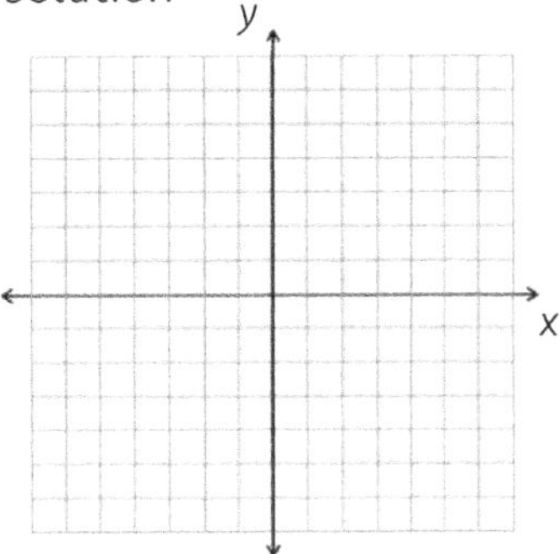

c. two solutions

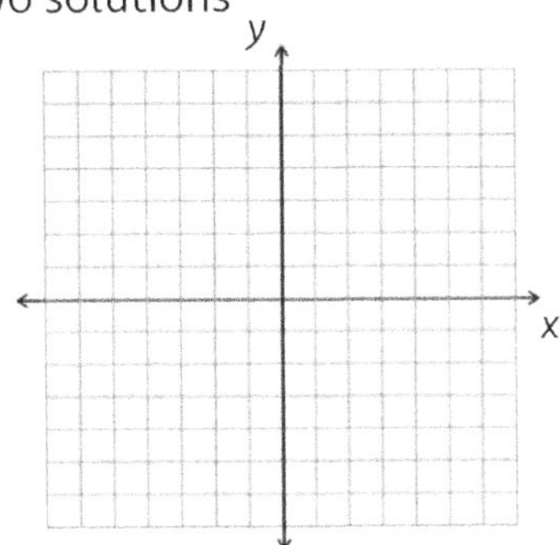

d. three solutions

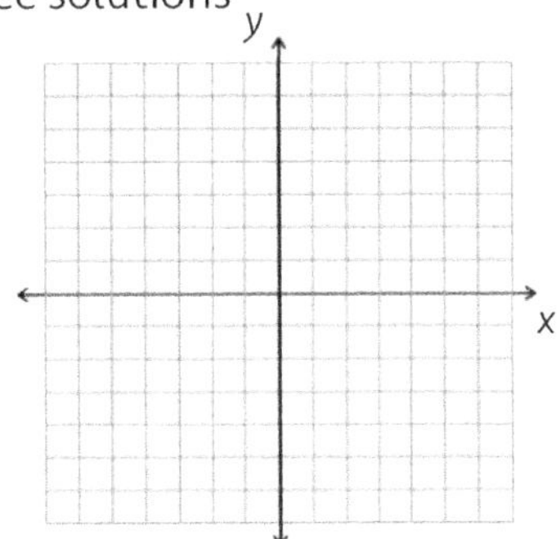

80. ★Solve the system shown below and verify your results by using a graphing application.

$$\begin{cases} x^2 + y^2 = 9 \\ y = x^2 - 3 \end{cases}$$

81. Solve the system shown below and verify your results by using a graphing application.

$$\begin{cases} x^2 + y^2 = 4 \\ x^2 + y^2 = 16 \end{cases}$$

82. Explain why the circles in the previous scenario do not intersect.

83. A small bug stands on a branch and drops down into a pond below. A jumping spider sits on a different branch below the bug and times its jump perfectly to catch the bug in its mouth as the bug drops. The bug's path is modeled by the function $H(t) = -4.9t^2 + 3$, while the spider's path is modeled by the function $H(t) = -4.9t^2 + 4t + 1$. In these functions, H is the bug or spider's height, in meters, t seconds after it jumps.

a. If the spider jumps up at the same moment that the bug drops, how many seconds has the bug been falling when it lands in the spider's mouth?

b. How high above the ground is the spider when it catches the bug?

84. The Bottle Company produces one type of drinking bottle. There is a cost associated with producing the bottles, but the company overcomes these costs by selling the bottles at a high enough price. The graphs for the revenue, $R(x)$, and costs, $C(x)$, are shown below, where x represents the number of bottles that have been produced and sold in a given month. Assume that every bottle that is produced is also sold.

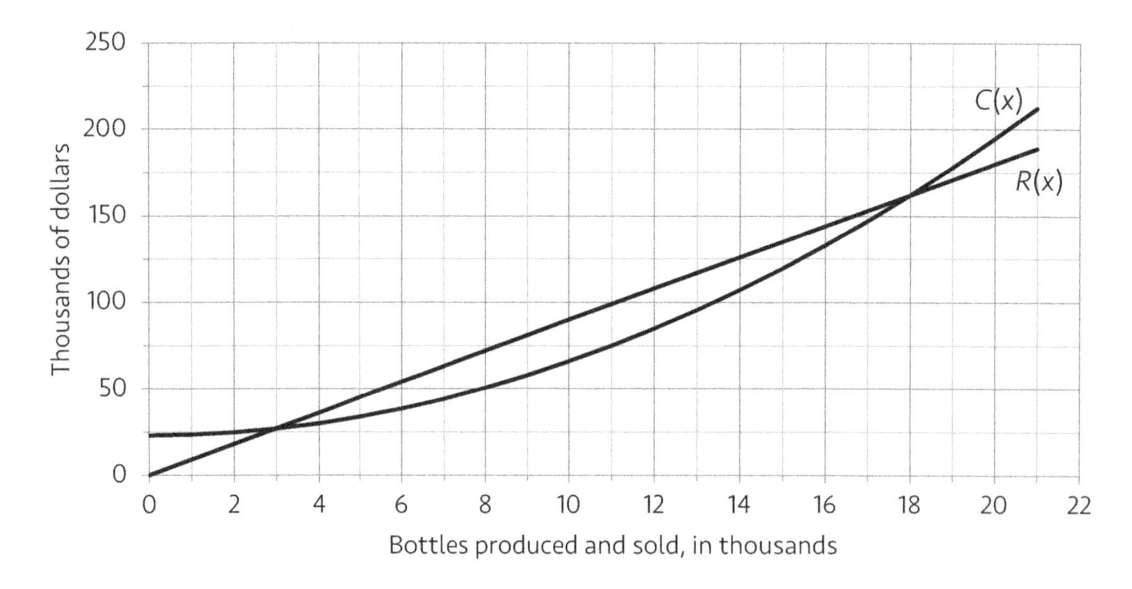

a. Use the graph to estimate the company's costs if they produce 8,000 bottles.

b. In order to sustain itself, a company needs to be profitable. Profit is defined as the revenue (money received) minus the costs (expenses). What is the company's profit if they produce 8,000 bottles? Make a rough estimate.

Use this page to record important ideas in the previous section or for any other writing that helps you learn the topics in this book.

Section 5

SYSTEMS WITH 3 VARIABLES

Two approaches that can be used to solve a system of equations with two unknowns are the Substitution Method and the Elimination Method. In the Substitution Method, you isolate one variable and make a substitution into another equation to replace that variable with an equivalent expression. In the Elimination Method, you combine two equations through addition or subtraction in such a way that one of the variables disappears, leaving behind a one-variable equation.

85. Solve the system of equations using the Substitution Method. Then start over and solve the system using the Elimination Method.

$$2x-3y=2$$
$$2y-x=1$$

If you start with a system of equations with three variables, you will need to manipulate the equations until you create a two-variable system. Once you accomplish this, the rest of the process is manageable because you already know how to solve two-variable systems.

86. The first thing you need to practice is manipulating 3-variable equations. In each equation shown, isolate the requested variable.

a. Isolate x.

$x + y + 2z = 7$

b. Isolate y.

$2x + y - 3z = 5$

c. Isolate z.

$x - 2y + z = 0$

87. Next, make a substitution as requested. Notice that the substitution converts a 3-variable equation into a 2-variable equation.

a. Substitute $z = 2y - x$ into the equation $x + y + 2z = 7$. Simplify the result.

b. Substitute $z = 2y - x$ into the equation $2x + y - 3z = 5$. Simplify the result.

88. You should now be able to solve the 3-variable system below, using the Substitution Method. Since z is isolated in the first equation, you can replace z with $2y - x$ in both of the other two equations to convert both of them to 2-variable equations.

$$z = 2y - x$$
$$x + y + 2z = 7$$
$$2x + y - 3z = 5$$

89. Use Substitution to convert the system to a 2-variable system. Do not solve the system. (To do this, select one equation and isolate a variable of your choice in that equation. Then make a substitution for the same variable in both of the other two equations. After you simplify each result, you should arrive at 2 equations with the same two variables. For example, if one equation contains x and y and the other contains y and z, you do not have a 2-variable system.)

$$x - 2y + 3z = 7$$
$$2x + y + z = 4$$
$$-3x + 2y - 2z = -10$$

90. Now that you have converted the previous 3-variable system into a 2-variable system, use what you already know about solving systems to finish the work that you started. Your remaining work should lead you to find the value of all 3 variables.

91. Solve the 3-variable system of equations using the Substitution Method.

$$2x-4y+2z=0$$
$$3x+3y+6z=21$$
$$2x+y-3z=5$$

92. Now use what you know about the Elimination Method and apply that to 3-variable equations. In each scenario, elimination can use two 3-variable equations to make one 2-variable equation.

a. Eliminate x.

$$x-2y+z=0$$
$$x+y+2z=7$$

b. Eliminate y.

$$2x+y-3z=5$$
$$x-2y+z=0$$

c. Eliminate z.

$$x+y+2z=7$$
$$2x+y-3z=5$$

93. Use Elimination to turn the given system into a 2-variable system. Do not solve the system. (To do this, select two equations and eliminate one of the variables. Then take another pair of equations and eliminate that same variable. After you do this, you should arrive at 2 equations with the same two variables.)

$$x-2y+3z=7$$
$$2x+y+z=4$$
$$-3x+2y-2z=-10$$

94. Now that you have converted the previous 3-variable system into a 2-variable system, use what you already know about solving systems to finish the work that you started. Your remaining work should lead you to find the value of all 3 variables.

95. Solve the 3-variable system of equations using the Elimination Method.

$$x-2y+z=0$$
$$x+y+2z=7$$
$$2x+y-3z=5$$

Have you wondered yet what the solution to a 3-variable system represents? An example from Geometry may help. The graph of a 2-variable equation is a line. When you graph two lines, if they intersect, they will intersect at a point. The graph of a 3-variable equation is a plane (imagine a sheet of paper extending infinitely in all directions). When you graph 3 planes, if they intersect at a common point, the coordinates can be written as (x, y, z).

96. Solve the 3-variable system of equations using the Substitution Method.

$$x+y+z=2$$
$$3x+3y+3z=14$$
$$x-2y+z=4$$

97. Solve the 3-variable system of equations using the Elimination Method.

$$x+y-2z=5$$
$$x+2y+z=8$$
$$2x+3y-z=13$$

98. It can be difficult to picture how separate planes can intersect at infinitely many points. One way to see how this can happen is to think about pages in a thin book. If you open a thin book and hold it open with two hands, you can pretend that each hand is holding a separate plane. Why does this scenario help you to see that separate planes can intersect at infinitely many points?

99. ★Solve the 3-variable system of equations using any combination of methods.

$$2x+y=-2$$
$$-x+3y-4z=-26$$
$$5x-6y+z=17$$

100. A business in Germany mails a total of 3,000 fundraising letters to 3 different countries, England, Poland, and France. They pay postage costs of $1.25 for each letter sent to England, $1.40 per letter sent to Poland, and $1.75 per letter sent to France. The number of letters sent to France was the same as the number of letters sent to England. The total postage costs were $4,420.

a. Write the equations that show the relationships described in this scenario.

b. How many letters does the business send to each country?

101. When you graduate from college, your income from summer jobs over the years has allowed you to accumulate $15,000 in savings. You take that money and invest it in 3 different accounts. After one year, the 3 accounts earn 2%, 5%, and 8% in interest, respectively. As a result of these interest rates, the total value of your investments increases by $720 that year. You invested $4000 more in the account that earned 5% than in the account that earned 2%.

a. Write the equations that show the relationships described in this scenario.

★b. How much money did you invest in each account?

Use this page to record important ideas in the previous section or for any other writing that helps you learn the topics in this book.

Section 6

FINDING THE EQUATION FOR A PARABOLA, GIVEN 3 POINTS

Earlier, you figured out how to find the equation of a line if you know two points on that line. In another lesson, you learned how to solve systems that involved two (and most recently, three) variables. You can use your knowledge of these concepts to find the equation of a parabola.

102. Start with a linear function that contains the points (1, −3) and (4, 6).

a. Describe how you would find the equation of this function, in Slope-Intercept Form.

b. There is another strategy for finding an equation of a line that involves creating a system of equations.

Equation 1. Since (1, −3) is a point on the line, you can replace x with 1 and y with −3 in the equation $y = Mx + B$. This creates the equation $-3 = M(1) + B$.

Equation 2. Since (4, 6) is also a point on the line, you can also replace x with 4 and y with 6. This creates the equation $6 = M(4) + B$.

c. Now you have two equations that both contain the M and B that you are trying to find. Solve this 2-variable system to find the equation of the line.

103. Now suppose there is a quadratic function that contains the points (2, 1), (3, 4), and (4, 9).

a. Plot the points in the graph to the right.

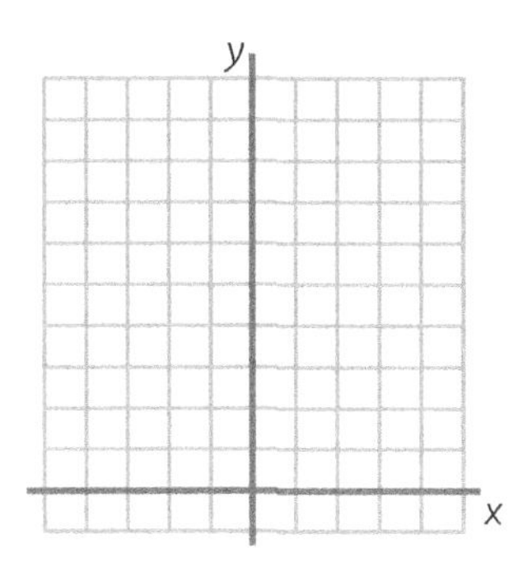

b. These points form part of a parabola and it is possible to use these points to find the equation of this parabola. To start with, write down the Standard Form for the equation of a parabola.

If you know A, B, and C in $y = Ax^2 + Bx + C$, then you have a complete equation. You can find these values using a system of equations like you did with the linear function in the previous scenario.

104. As a reminder, this parabola passes through the points (2, 1), (3, 4), and (4, 9). To start with, since (2, 1) is a point on this parabola, you can replace x with 2 and y with 1 to get the equation $1 = A(2)^2 + B(2) + C$. This can be simplified to become $1 = 4A + 2B + C$.

a. Plug in the other two known points, (3, 4), and (4, 9), and write the two equations that you get when you do this.

b. You now have 3 equations containing the variables A, B, and C. Solve this system of equations. Then write the equation of the parabola in Standard Form.

c. Once you have the equation, fill in the missing values in the coordinates below and plot these points to get a more complete graph of this parabola.

(−2, ____), (−1, ____), (0, ____), (1, ____)

105. Consider a new group of three points: (1, 8), (2, 7), (3, 4). Begin the process of finding the equation of the parabola, $y = Ax^2 + Bx + C$, that contains these points. Just as you did in the previous scenario, determine the three equations that contain A, B, and C. Do **not** solve this system of equations. Just write them.

106. A parabola passes through the points (−3, 9), (6, 0), and (−7, 1). Begin the process of finding the equation of the parabola, $y = Ax^2 + Bx + C$, that contains these points. Use the 3 points to write the three equations that contain A, B, and C. Do **not** solve this system of equations. Just write them.

107. A systems of three equations is shown. What are the coordinates of the 3 points that the system of equations represents? Do **not** solve the system. Instead, only write the 3 points.

a. $$\begin{aligned} -2 &= A - B + C \\ 7 &= 4A + 2B + C \\ 3 &= 16A + 4B + C \end{aligned}$$

b. $$\begin{aligned} 9 &= 16A - 4B + C \\ 7 &= 36A - 6B + C \\ 1 &= 0A + 0B + C \end{aligned}$$

108. Each system in the previous scenario represents a parabola. Find the equation for each parabola.

109. Find the equation for the parabola that passes through the following points.

$(4, -4)$, $(5, -3)$, $(6, 0)$

110. Consider the table below.

x	1	2	3	4	5
y	8	7	4	-1	

a. Identify a pattern in the table above and use the pattern to find the missing term.

b. The ordered pairs are points from a parabola. Find the equation of this parabola.

111. Determine the equation of the parabola shown.

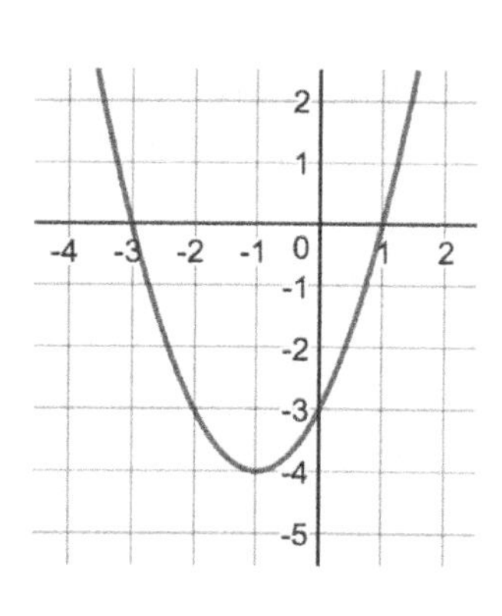

112. A goalie kicks a soccer ball high in the air to move the ball to the other side of the field. After 1 second the height of the ball is 80 feet. After 1.5 seconds the height of the ball is 108 feet. After 2 seconds the height of the ball is 128 feet.

a. After how many seconds will the ball hit the ground?

b. What is the maximum height of the ball during its trajectory?

113. There is a ramp on the flat roof of a building and the bottom of the ramp is at the edge of the roof, 144 feet above the ground. When a marble is rolled down the ramp it is launched outward and follows a parabolic arc as it falls down to the ground below. The marble does not follow a parabolic path until the moment it is launched. One second after it is launched, the marble is 128 feet above the ground. After two seconds, it is 80 feet above the ground.

a. The previous scenario gave you 3 distinct ordered pairs, but this scenario seems to give you only two ordered pairs. If you need three points to determine a parabola, what is the third ordered pair in this scenario?

★b. After how many seconds will the marble hit the ground?

Use this page to record important ideas in the previous section or for any other writing that helps you learn the topics in this book.

Section 7
CUMULATIVE REVIEW

114. Write out The Quadratic Formula.

115. Draw a very basic sketch of a parabola that has each of the following characteristics.

a. two x-intercepts
opens downward

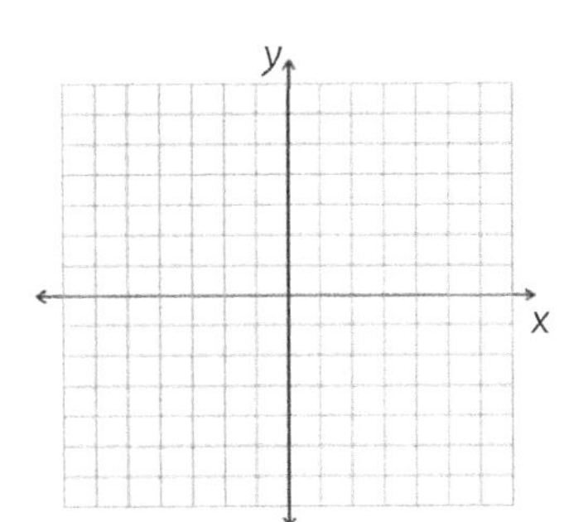

b. one x-intercept
opens upward

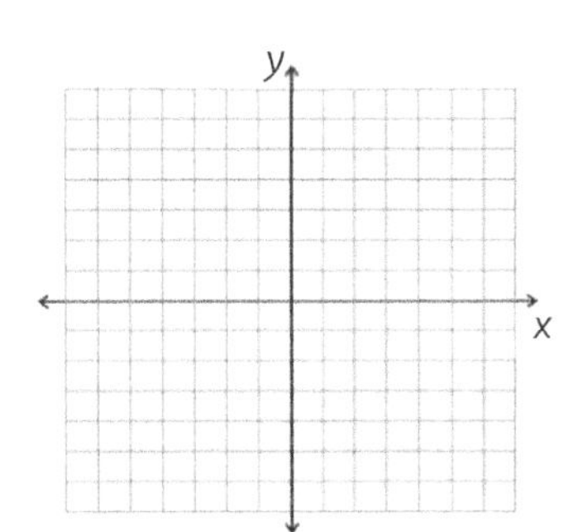

c. no x-intercepts
opens downward

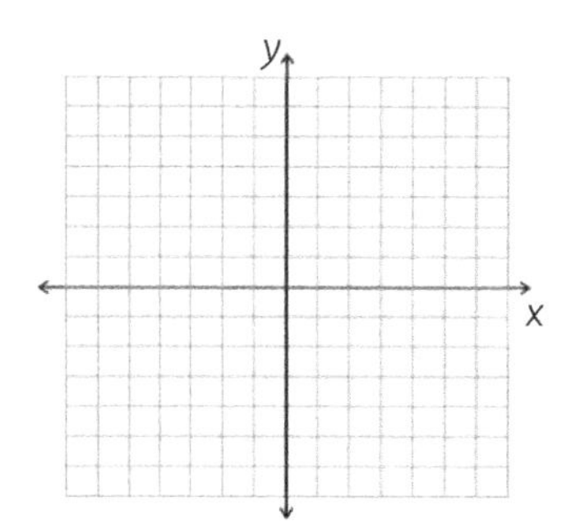

116. How can you use a parabola's equation to determine if a parabola will open upward or downward?

117. Determine the x-intercepts of the function $f(x) = 2x^2 - 3x + 5$.

118. If the vertex of a parabola is located at (4, −2) and one of the x-intercepts is (−1, 0), where is the other x-intercept located?

119. A rocket is launched from a position on the ground and its trajectory follows a path modeled by the equation $H(t) = -16t^2 + 80t$, where H is the height of the rocket, in feet, t seconds after it is launched.

a. What is the maximum height of the rocket?

b. For how many total seconds is the rocket in the air?

120. Graph the function $f(x) = -x^2 + 2x + 2$. Plot at least 7 points, including the vertex, the x-intercept(s), and the y intercept.

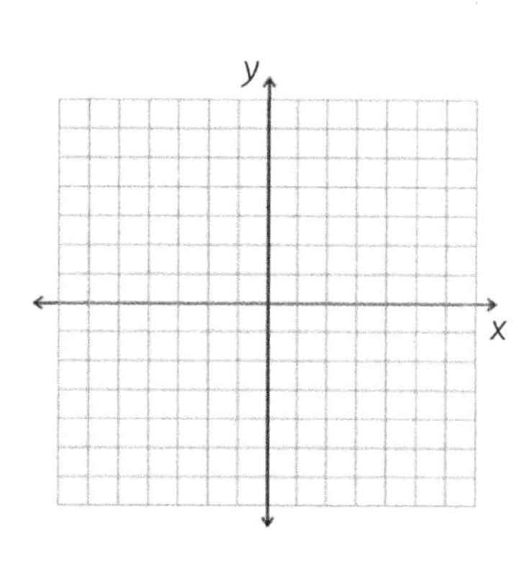

121. Solve the equation.

$$150 = 125 + 25t^2$$

122. ★Calculate the exact area of the square formed by the system of inequalities shown.

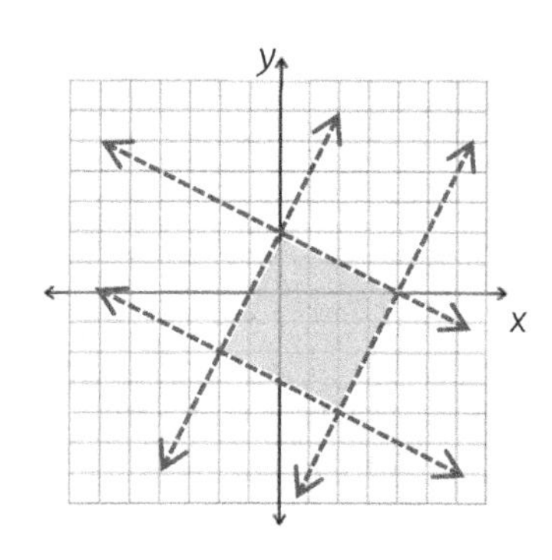

123. ★What is the exact area of the circle shown?

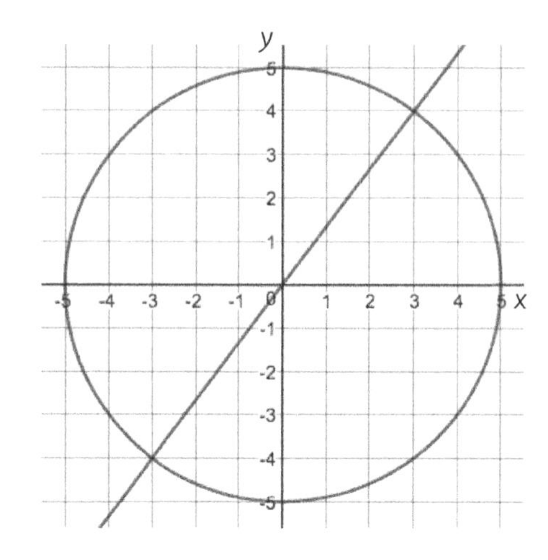

Use this page to record important ideas in the previous section or
for any other writing that helps you learn the topics in this book.

Section 8
ANSWER KEY

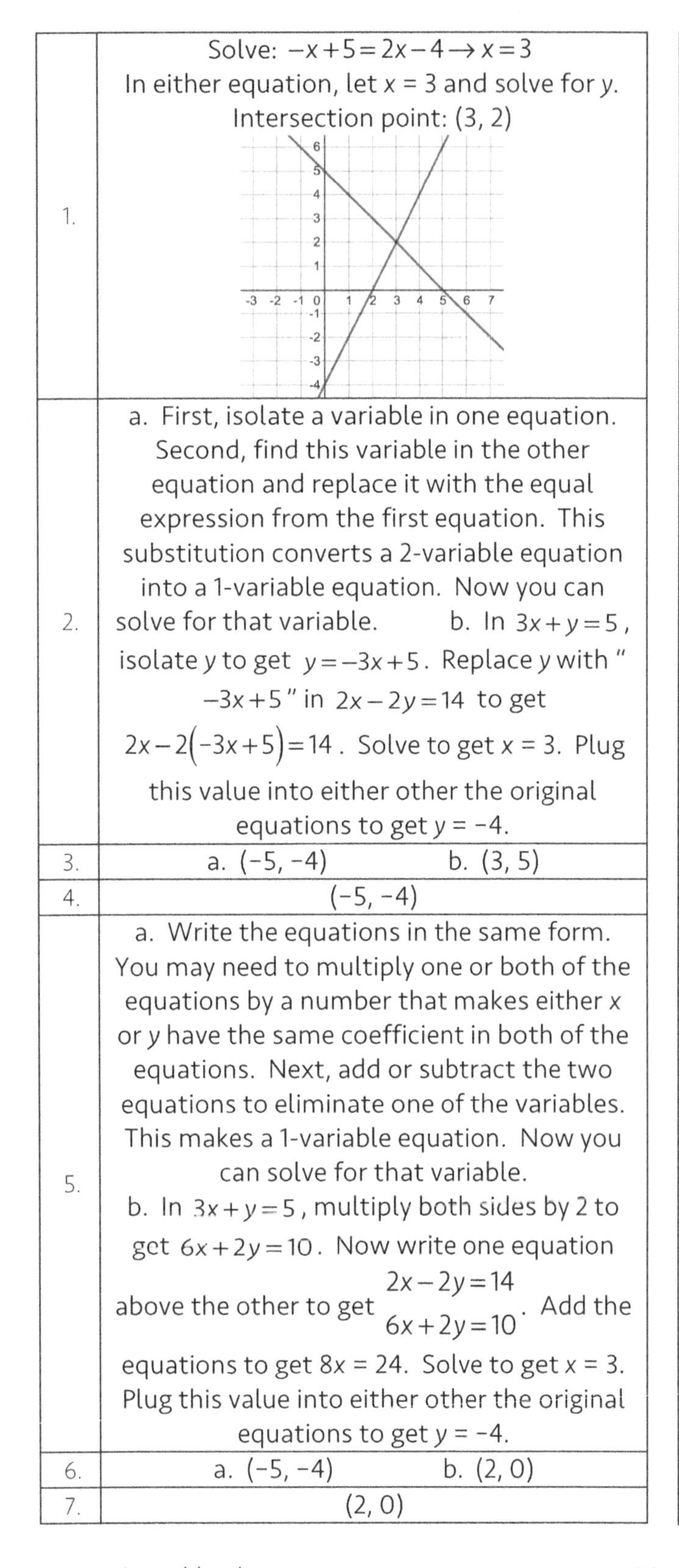

1.	Solve: $-x+5=2x-4 \rightarrow x=3$ In either equation, let x = 3 and solve for y. Intersection point: (3, 2)
2.	a. First, isolate a variable in one equation. Second, find this variable in the other equation and replace it with the equal expression from the first equation. This substitution converts a 2-variable equation into a 1-variable equation. Now you can solve for that variable. b. In $3x+y=5$, isolate y to get $y=-3x+5$. Replace y with "$-3x+5$" in $2x-2y=14$ to get $2x-2(-3x+5)=14$. Solve to get x = 3. Plug this value into either other the original equations to get y = -4.
3.	a. (-5, -4) b. (3, 5)
4.	(-5, -4)
5.	a. Write the equations in the same form. You may need to multiply one or both of the equations by a number that makes either x or y have the same coefficient in both of the equations. Next, add or subtract the two equations to eliminate one of the variables. This makes a 1-variable equation. Now you can solve for that variable. b. In $3x+y=5$, multiply both sides by 2 to get $6x+2y=10$. Now write one equation above the other to get $\begin{matrix} 2x-2y=14 \\ 6x+2y=10 \end{matrix}$. Add the equations to get $8x$ = 24. Solve to get x = 3. Plug this value into either other the original equations to get y = -4.
6.	a. (-5, -4) b. (2, 0)
7.	(2, 0)

8.	$y=-\frac{1}{3}x+2$
9.	Line 1: Slope-Intercept Form: $y=-x+1$ Standard Form: $x+y=1$ Line 2: Slope-Intercept Form: $y=2x+5$ Standard Form: $-2x+y=5$
10.	$\left(-\frac{4}{3}, 2\frac{1}{3}\right)$
11.	Two points. You can use them to find the slope and then the y-intercept.
12.	Solve the system: $\begin{cases} y=x-2 \\ y=-\frac{5}{7}x+5 \end{cases}$ Intersection: $\left(\frac{49}{12}, \frac{25}{12}\right)$ or $\left(4\frac{1}{12}, 2\frac{1}{12}\right)$
13.	a. $2\frac{1}{12}$ units b. $4\frac{1}{12}$ units
14.	a. Draw a line that passes through (3, 2). b. Draw a line that runs parallel to the one already shown.
15.	a. original: $y=\frac{1}{3}x+1$ new (many possible): $y=-\frac{1}{3}x+3$, $y=\frac{2}{3}x$, $y=x-1$, $y=-x+5$, etc... b. original: $y=-\frac{1}{3}x+2$ new: $y=-\frac{1}{3}x+\text{anything}$
16.	They do not intersect.
17.	The lines are parallel. They have the same slope.
18.	The lines intersect infinitely many times. The equations are the same.

#	Answer
19.	a. slopes will be opposite reciprocals b. original: $y=\frac{2}{3}x-2$ new: $y=-\frac{3}{2}x-4$ 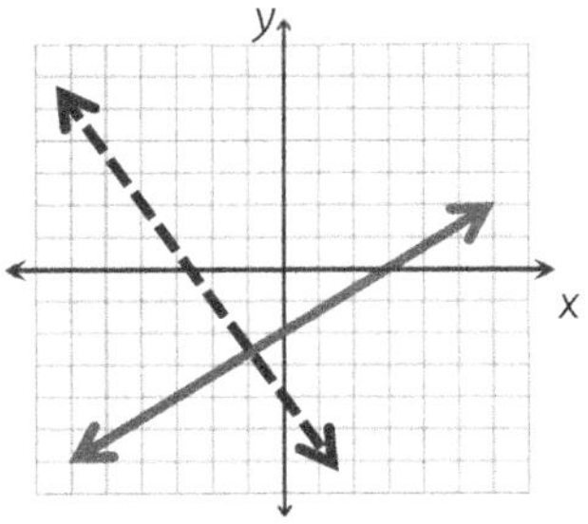
20.	$y=-\frac{2}{3}x-4$ 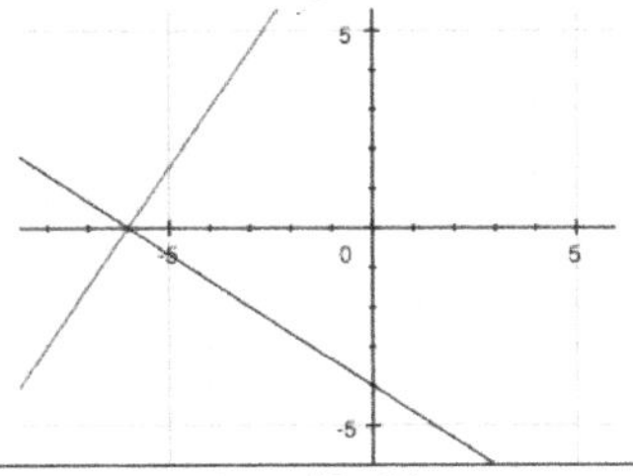
21.	Yes, (2, 3).
22.	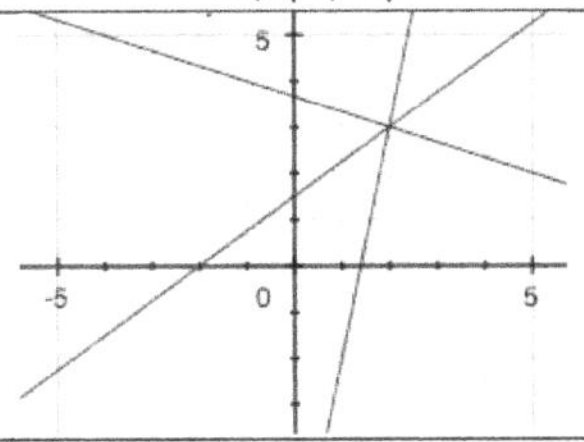
23.	$x = 10$ and $y = -3$ so $xy = -30$
24.	Area = $\frac{1}{2}(8)(6)=24$ units^2
25.	Area = $\frac{1}{2}(8)(2)=8$ units^2
26.	a. $10\cdot4=40$ cm^2 b. $\frac{1}{2}(10\cdot4)=20$ in.^{2} c. $\frac{10\cdot4}{2}=20$ ft^2
27.	a. Around 10 or 11 units^2 b. Intersection point: $(-1.6, 3.6)$ Area = $\frac{1}{2}(6)(3.6)=10.8$ units^2
28.	a. Around 14 or 15 units^2 b. Intersection point: $\left(1\frac{1}{5}, 2\frac{1}{5}\right)$ Area = $\frac{1}{2}(7)\left(4\frac{1}{5}\right)=14\frac{7}{10}$ units^2
29.	(0.6, 5.8)
30.	a. Washington = 30°; Beijing = 45° b. 5 hours (solve $-2h+45 = h+30$)
31.	a. the first resort: $2,075 b. 10 days
32.	Pizza = $15.75 Hamburger = $3.25
33.	a. 11 ounces (it would cost $3.65) b. Yogoberry: 30 cents per ounce
34.	Equations: $r + a = 5$ and $4.25r + 7.50a = 25.15$ raisins: 3.8 lbs almonds: 1.2 lbs
35.	Equations: $n + d = 580$ and $0.05n + 0.10d = 41.80$ 324 nickels, 256 dimes b. $39.71
36.	A + U + G = 540; G=1.25(A+U) Grandfather: $300
37.	Equations: Salary = 11,500 + 0.08x and Salary = 25,000 + 0.04x a. Maria: Alpha; Rose: Beta b. $337,500 c. M: $43,660; R: $37,860; L: $38,500
38.	a. solve $17.2 - 1.2h = 12.8 - 0.8h$; $h = 11$ hours b. 5:00am
39.	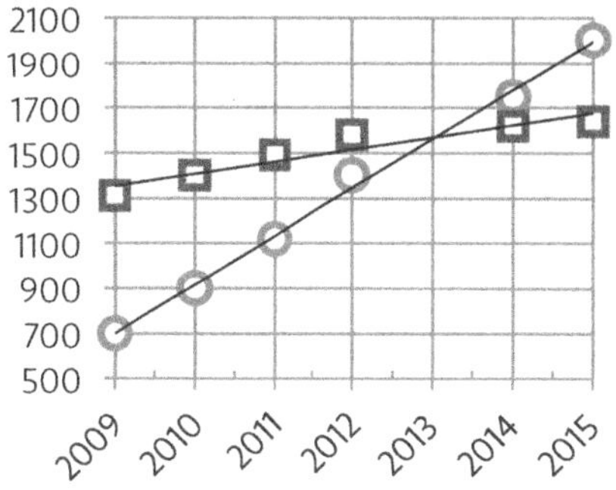
40.	The slope of each line represents the average increase in users per year desktop growth: 60–70 million per year mobile growth: ≈200 million per year
41.	a. one b. the lines are parallel c. the two lines are the same, so they intersect infinitely many times
42.	Infinitely many points 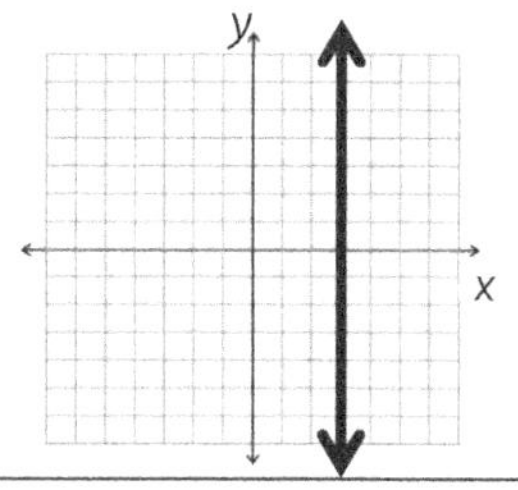
43.	Infinitely many points 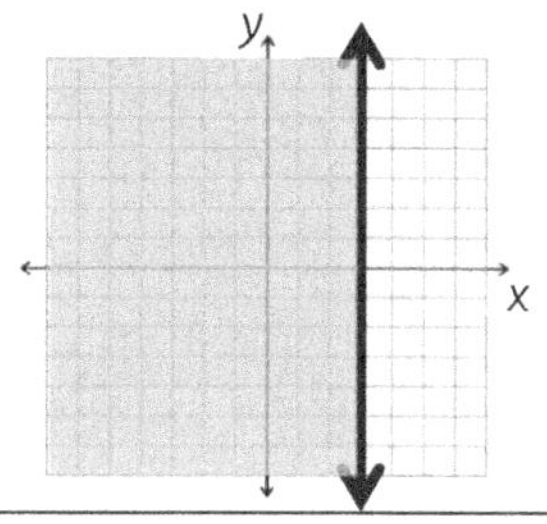

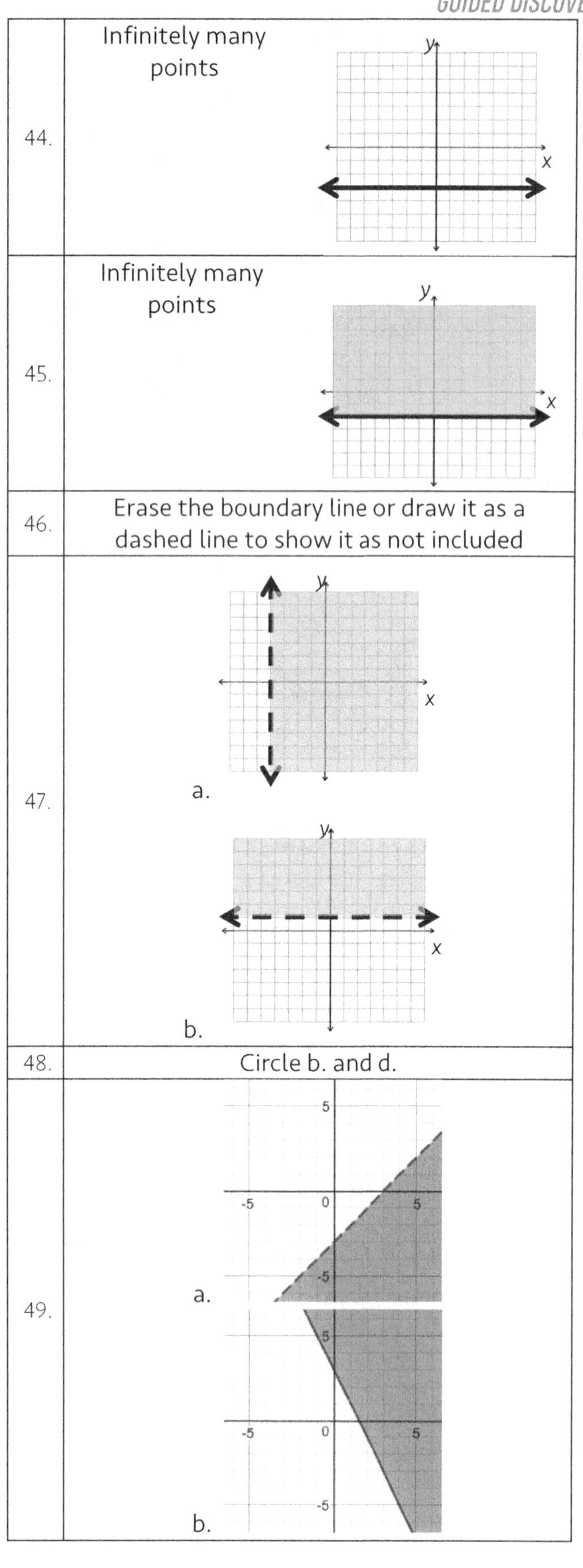

44.	Infinitely many points
45.	Infinitely many points
46.	Erase the boundary line or draw it as a dashed line to show it as not included
47.	a. b.
48.	Circle b. and d.
49.	a. b.

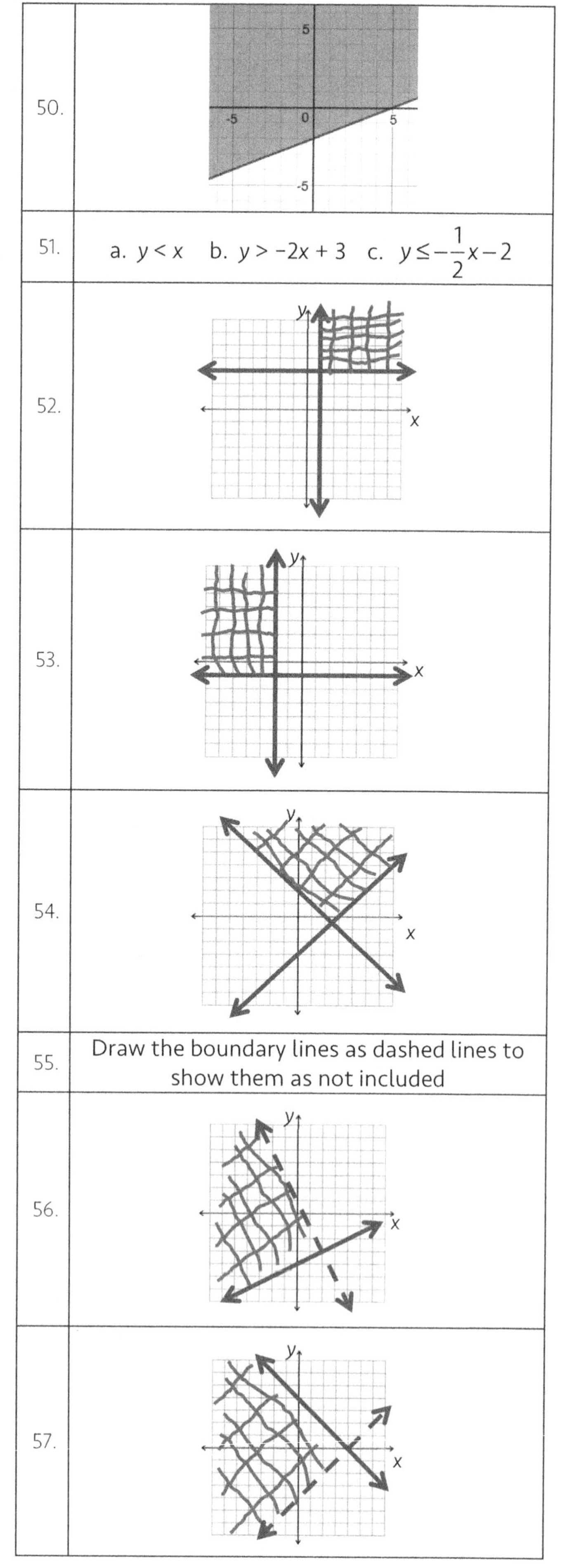

50.	
51.	a. $y < x$ b. $y > -2x + 3$ c. $y \le -\frac{1}{2}x - 2$
52.	
53.	
54.	
55.	Draw the boundary lines as dashed lines to show them as not included
56.	
57.	

58.	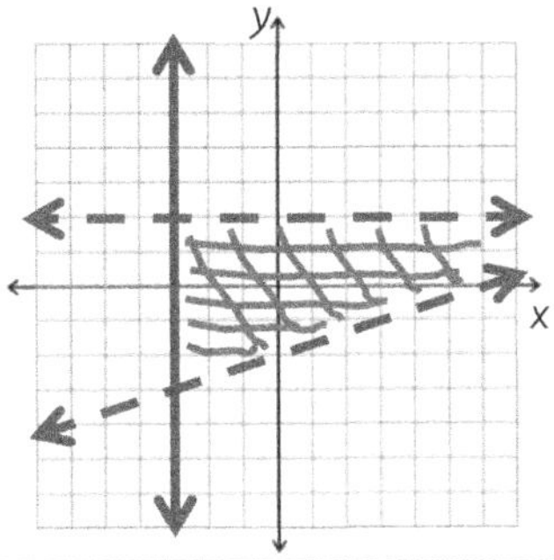
59.	Area = $\frac{1}{2}(5)(15)$ = 37.5 units^2
60.	$y \le -\frac{1}{3}x+3,\ y > -\frac{3}{2}x-1,\ y \ge 4x-3$
61.	a. Intersection point: $\left(\frac{60}{11}, \frac{95}{11}\right)$ b. $y > \frac{3}{5}x-4,\ y > -\frac{2}{3}x+5,$ $y < -\frac{4}{5}x+13,\ y < \frac{2}{3}x+5$
62.	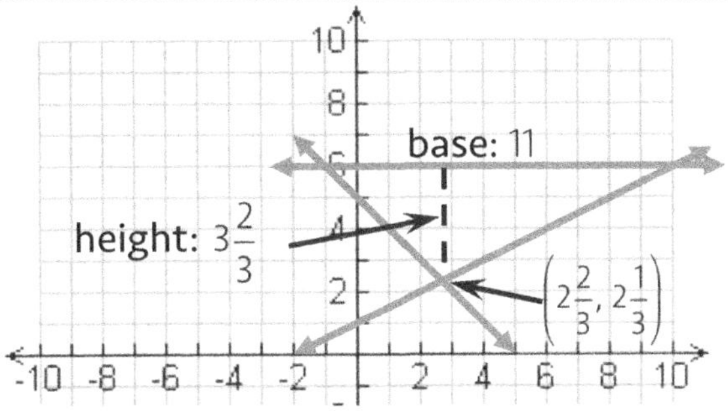 base: 11 height: $3\frac{2}{3}$ $\left(2\frac{2}{3}, 2\frac{1}{3}\right)$ Area = $\frac{1}{2}(11)\left(\frac{11}{3}\right) = \frac{121}{6} = 20\frac{1}{6}$ = units^2
63.	a. $3H + 5P \ge 15$, $3H + 5P \le 30$ b. shaded region is between two parallel lines c. – d. look for lattice points (integer-only coordinates) e. 29
64.	$x+2y<4 \quad y>-\frac{1}{2}x-3$ $y>2x-8 \quad y<2x+2$
65.	a. $x^2+8x+16$ b. f^2-2f+1 c. y^4-20y^2+100

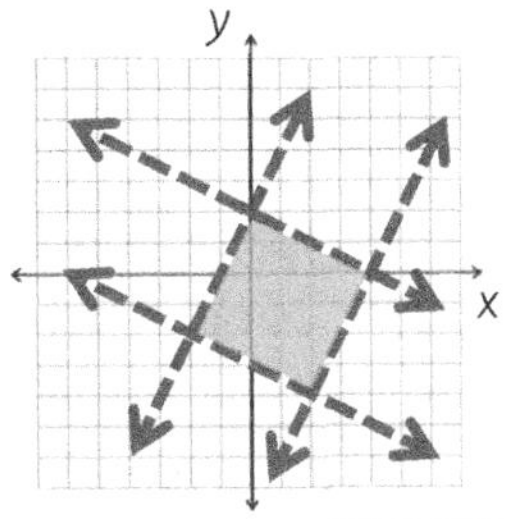

66.	$\frac{16}{25}x^2$
67.	$-(-3)^2+2(-3)+7 \rightarrow -9-6+7 \rightarrow -8$
68.	The solution is the point where the graphs of a system of equations intersect.
69.	2 solutions: (-3, -4) and (3, 4)
70.	(-4, -1), (0, 1), (4, 3)
71.	(-4, -7) and (3, 0) Insert each x-value into either equation to find the y-value that goes with the x-value.
72.	(2, -1) and (-2, 3) Solve: $-x^2-x+5=x^2-x-3 \rightarrow x=2$ or -2 Insert each x-value into either equation to find the y-value that goes with the x-value.
73.	(-1, 2), (0, 1), (1, 0) One option: solve $x^3-2x+1=-x+1$ x = 0, 1 or -1 Plug in each x-value to solve for y.
74.	a. $y=-\frac{1}{4}x^2+4$ b. A = 4
75.	a. (-1, -5), (5, 7); solve $x^2-2x-8=2x-3$ $\rightarrow x^2-4x-5=0 \rightarrow (x-5)(x+1)=0$ $\rightarrow x=5$ or -1 b. (2, 1) ; solve $x^3-2x^2+x-1=7+x-2x^2$ $x^3-8=0 \rightarrow x^3=8 \rightarrow x=2$
76.	(-4, -3), (3, 4); solve $x^2+(x+1)^2=25$ $x^2+(x^2+2x+1)=25 \rightarrow 2x^2+2x-24=0$ $2(x+4)(x-3)=0 \rightarrow x=-4$ or 3
77.	a. 2 solutions b. (-2, 3), (3, -2)
78.	x^2 = 36
79.	a. b. c. d. 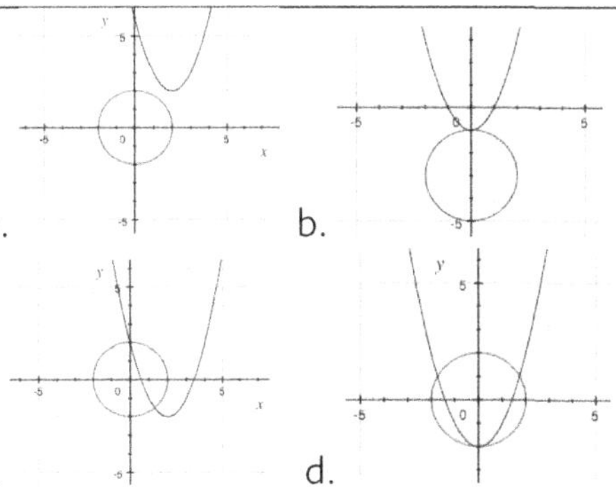
80.	$(0,-3), (\sqrt{5},2), (-\sqrt{5},2)$ solve: $x^2+(x^2-3)^2=9 \rightarrow x^2+(x^4-6x^2+9)=9$ $x^4-5x^2=0 \rightarrow x^2(x^2-5)=0 \rightarrow x=0, \sqrt{5}, -\sqrt{5}$
81.	No solution. The equations represent two circles that do not intersect.

82.	The circles have the same center at (0,0), but they do not have the same radius length.
83.	a. 0.5 seconds solve $-4.9t^2+3=-4.9t^2+4t+1$ b. 1.775 meters
84.	a. \$50,000 b. Slightly less than \$25,000
85.	(7, 4)
86.	a. $x = 7 - y - 2z$ b. $y = 5 - 2x + 3z$ c. $z = -x + 2y$
87.	a. $x + y + 2(2y - x) = 7 \rightarrow -x + 5y = 7$ b. $2x + y - 3(2y - x) = 5 \rightarrow 5x - 5y = 5$
88.	Use substitution or elimination to solve the system $-x + 5y = 7$ and $5x - 5y = 5$. $x = 3,\ y = 2,\ z = 1 \rightarrow (3, 2, 1)$
89.	–
90.	$x = 2,\ y = -1,\ z = 1 \rightarrow (2, -1, 1)$
91.	$x = 3,\ y = 2,\ z = 1 \rightarrow (3, 2, 1)$
92.	a. $-3y - z = -7$ b. $5x - 5z = 10$ c. $7x + 5y = 31$
93.	–
94.	$x = 2,\ y = -1,\ z = 1 \rightarrow (2, -1, 1)$
95.	(3, 2, 1)
96.	no solution (two of the planes are parallel)
97.	infinite solutions (all 3 planes intersect along a common line)
98.	As you hold open a thin book, the separate planes intersect at the binding of the book. The binding is an edge so you can pretend it is a line. Since the separate planes both intersect at that line, and a line contains infinitely many points, the two planes intersect at an infinite number of points.
99.	(0, −2, 5)
100.	a. Equation 1: $E + P + F = 3{,}000$ Equation 2: $1.25E + 1.40P + 1.75F = 4{,}420$ Equation 3: $F = E$ b. 1100 letters to England and France 800 letters to Poland
101.	a. Equation 1: $a + b + c = 15{,}000$ Equation 2: $0.02a + 0.05b + 0.08c = 720$ Equation 3: $b = a + 4{,}000$ b. 2%: \$4000; 5%: \$8000; 8%: \$3000
102.	a. Use the Slope Formula to find the slope of the line, given the 2 points. Use one of the 2 points to find the *y*-intercept. c. M = 3 and B = −6 so $y = 3x - 6$

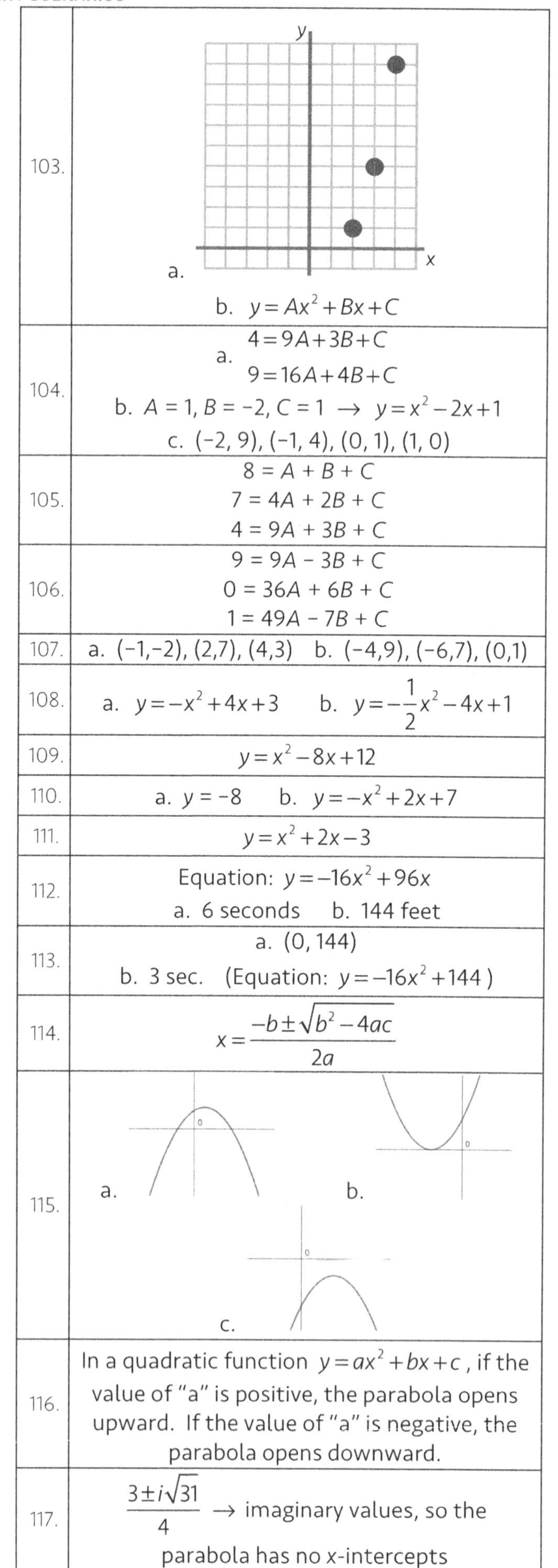

103.	a. 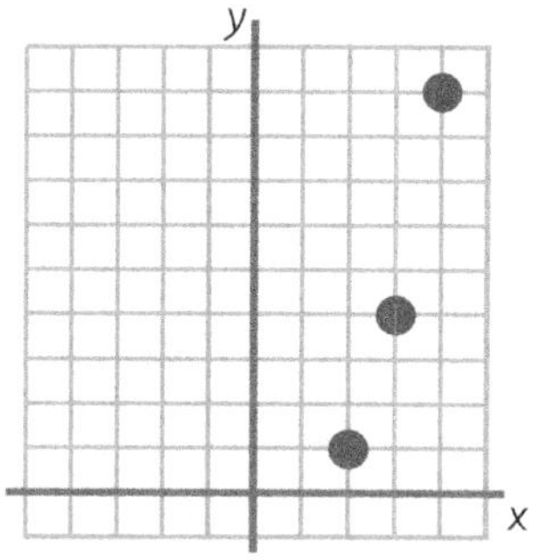 b. $y=Ax^2+Bx+C$
104.	a. $4=9A+3B+C$ $9=16A+4B+C$ b. $A = 1, B = -2, C = 1 \rightarrow y=x^2-2x+1$ c. (−2, 9), (−1, 4), (0, 1), (1, 0)
105.	$8 = A + B + C$ $7 = 4A + 2B + C$ $4 = 9A + 3B + C$
106.	$9 = 9A - 3B + C$ $0 = 36A + 6B + C$ $1 = 49A - 7B + C$
107.	a. (−1,−2), (2,7), (4,3) b. (−4,9), (−6,7), (0,1)
108.	a. $y=-x^2+4x+3$ b. $y=-\frac{1}{2}x^2-4x+1$
109.	$y=x^2-8x+12$
110.	a. $y = -8$ b. $y=-x^2+2x+7$
111.	$y=x^2+2x-3$
112.	Equation: $y=-16x^2+96x$ a. 6 seconds b. 144 feet
113.	a. (0, 144) b. 3 sec. (Equation: $y=-16x^2+144$)
114.	$x=\dfrac{-b\pm\sqrt{b^2-4ac}}{2a}$
115.	a. 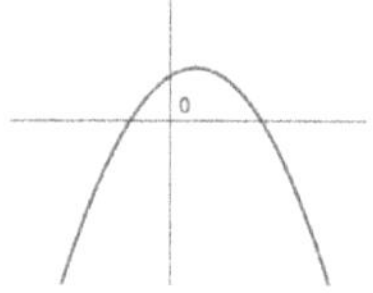 b. 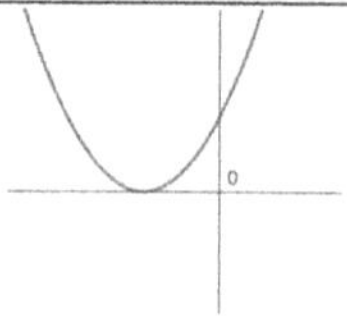 c. 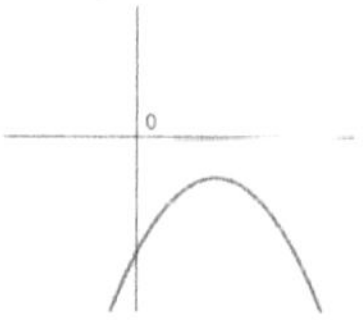
116.	In a quadratic function $y=ax^2+bx+c$, if the value of "a" is positive, the parabola opens upward. If the value of "a" is negative, the parabola opens downward.
117.	$\frac{3\pm i\sqrt{31}}{4} \rightarrow$ imaginary values, so the parabola has no *x*-intercepts

118.	(9, 0)
119.	a. 100 feet b. 5 seconds
120.	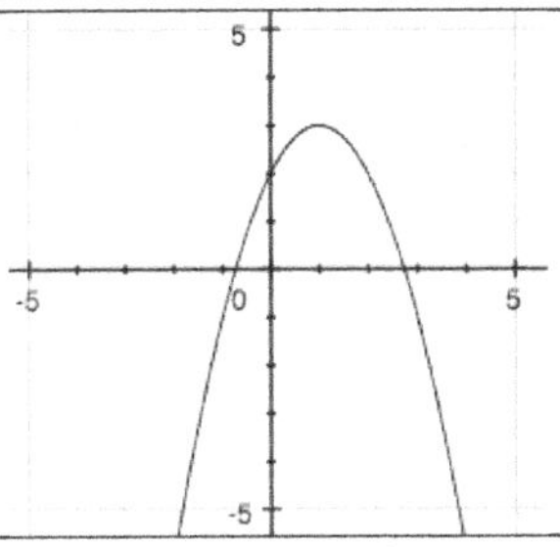
121.	$x = 1$ or -1

122.	$\left(2\sqrt{5}\right)^2 = 20$ units^2
123.	$\pi\left(5\right)^2 = 25\pi$

HOMEWORK & EXTRA PRACTICE SCENARIOS

As you complete scenarios in this part of the book, you will practice what you learned in the guided discovery sections. You will develop a greater proficiency with the vocabulary, symbols and concepts presented in this book. Practice will improve your ability to retain these ideas and skills over longer periods of time.

There is an Answer Key at the end of this part of the book. Check the Answer Key after every scenario to ensure that you are accurately practicing what you have learned. If you struggle to complete any scenarios, try to find someone who can guide you through them.

CONTENTS

Section 1

REVIEW GRAPHING SYSTEMS, SUBSTITUTION AND ELIMINATION

You were introduced to the topic of Systems of Equations in a previous lesson. As with many topics in mathematics, it is helpful to come back and take another look at a topic in order to strengthen your memory and extend what you know to learn more about that topic.

1. Without graphing, where do the two lines intersect? After you find this point, graph the lines to confirm that your intersection point is accurate.

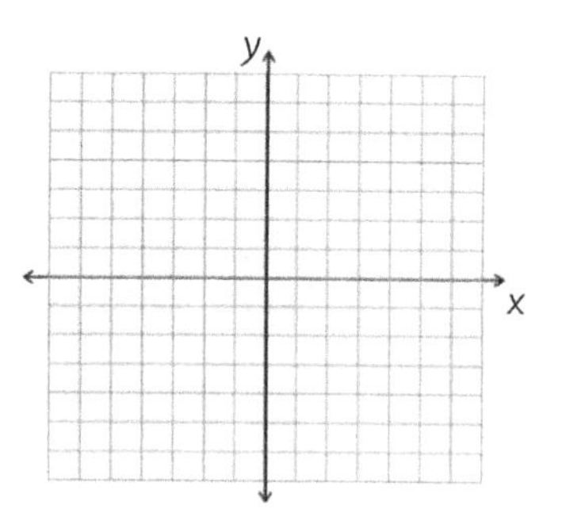

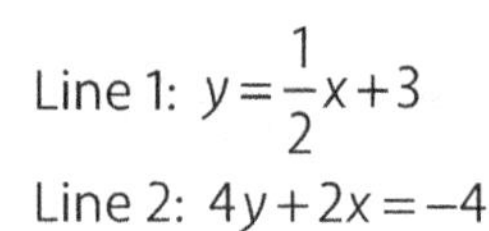

Line 1: $y=\frac{1}{2}x+3$

Line 2: $4y+2x=-4$

2. A system of equations is shown below. Why is it easier to solve the system using the Substitution Method instead of the Elimination Method?

$y=-2x+5$

$4x-y=7$

3. Solve the previous system of equations using the Substitution Method.

4. Solve the system of equations using the Substitution Method. As a reminder, you can make a substitution for either x or y. The goal is to change a 2-variable equation to make it contain only one variable.

$x=-\frac{1}{3}y+1$

$-2y+3x=-6$

5. Use the Substitution Method to solve each system of equations.

a. $x = 2y - 13$
$y - 3x = 4$

b. $y = \frac{1}{2}x - 9$
$2x - 8y = 20$

6. Solve the system of equations using the Substitution Method.

$-5y + 4x = 2$
$-3x + 6y = 12$

7. A system of equations is shown below. Why is it easier to solve the system using the Elimination Method instead of the Substitution Method?

$2x - 5y = 3$
$3x + 5y = 17$

8. Solve the previous system of equations using the Elimination Method.

9. Use the Elimination Method to solve each system of equations.

a. $6x - 11y = 1$
$3x + 11y = 17$

b. $y = 6x - 22$
$40x - 2y = 100$

10. Solve the system of equations using the Elimination Method. As a reminder, the equations can either be added or subtracted in order to eliminate one of the variables.

$5x + 3y = 11$
$-2x + 6y = 10$

11. Solve the system of equations using the Elimination Method. As a reminder, the equations can either be added or subtracted in order to eliminate one of the variables.

$-5y + 4x = 2$
$-3x + 6y = 12$

12. The x- and y-intercepts of both lines shown are integers. Estimate the coordinates of their intersection point and then determine the exact coordinates.

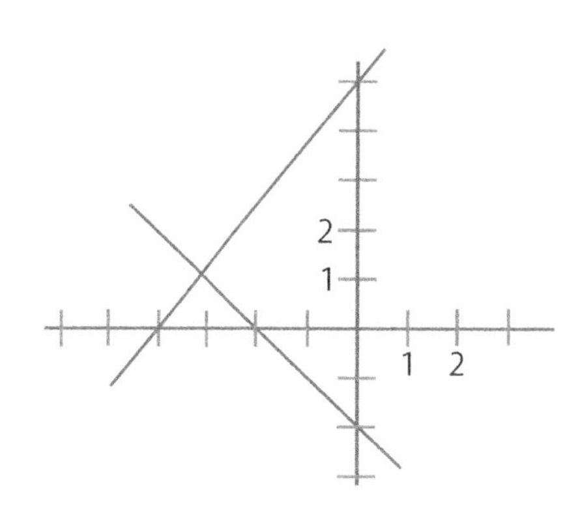

13. If you look at the two lines in the graph above and the x- and y-axes, you will notice that there are triangles formed by these converging lines and axes.

a. What is the height of the triangle that is formed by the x-axis and the two lines?

b. What is the height of the triangle that is formed by the y-axis and the two lines?

14. Without graphing them, where do the two lines intersect?

Line 1: $4x = 3y - 3$

Line 2: $6y + 2x = 6 + 10x$

15. Line 1 passes through the points (2, 1) and (−4, 4). Line 2 passes through the point (−3, −1). The system of equations formed by Line 1 and Line 2 has no solution.

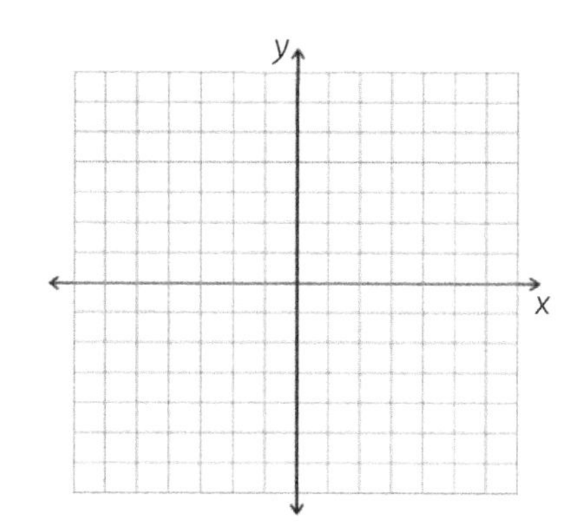

a. What is the equation of Line 2, in Slope-Intercept Form?

b. Draw both of the lines in the graph to the right.

16. Start with the line formed by the equation $2x+y=8$. Identify a second equation that forms two perpendicular lines with identical x-intercepts. Graph both of these lines to confirm that your second equation satisfies the stated condition.

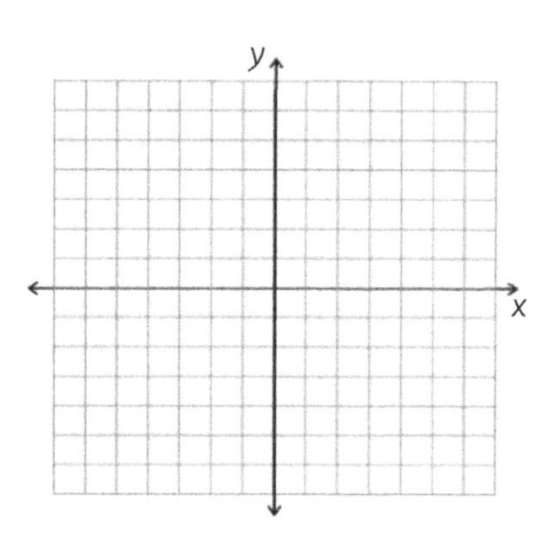

17. Consider three lines: $3x-2y=8$, $x-y=-1$, and $-x+4y=34$. Do they intersect at a single point? If so, find it. If not, explain why not.

18. Graph the previous three lines in such a way that their point of intersection is visible on the graph.

19. Calculate the area of each figure shown below.

a.

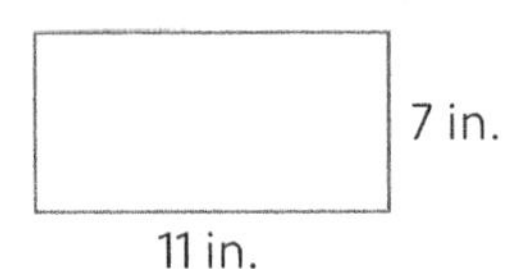

b.

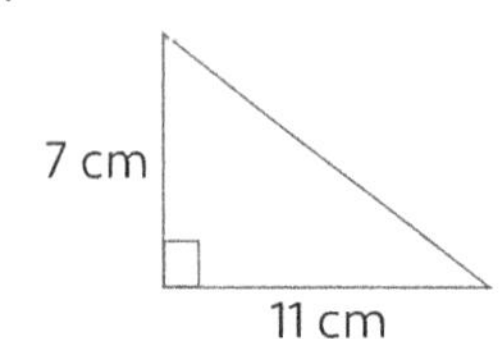

c.

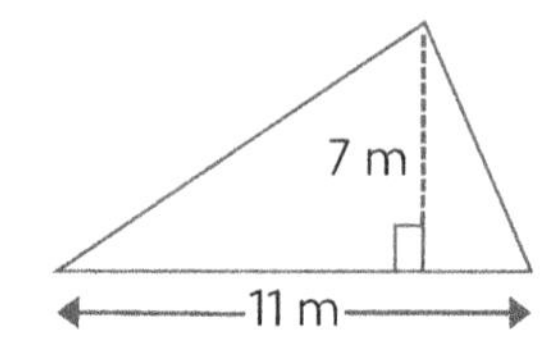

20. Calculate the area of the region that is contained between the y-axis and the graphs of the equations $2y=-x+18$ and $y=x$.

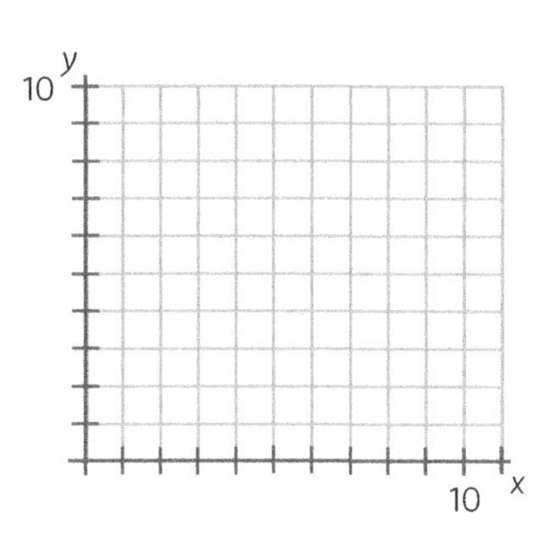

21. Graph the three equations below on the grid to the right.

$$\begin{cases} y=2x+6 \\ 2x+3y=12 \\ y=0 \end{cases}$$

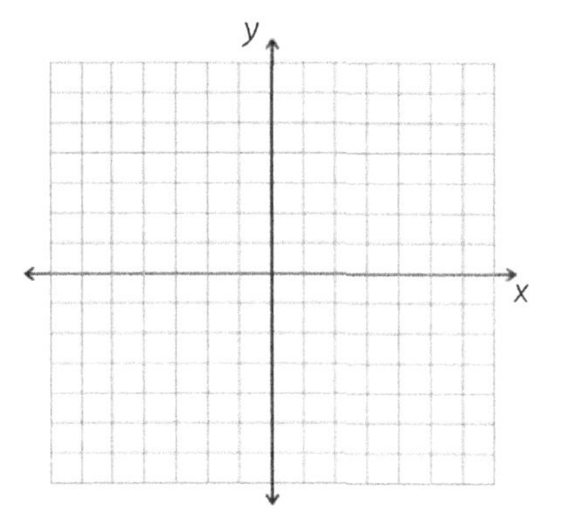

a. Estimate the area of the triangle formed by the three lines by counting the squares inside the triangle.

b. Calculate the exact area of the triangle enclosed by the three lines. Express your answer as a mixed number.

22. The figure shows the graphs of two lines, whose axis intercepts are integers. There is a third line drawn along the x-axis that forms a triangle with these other two lines.

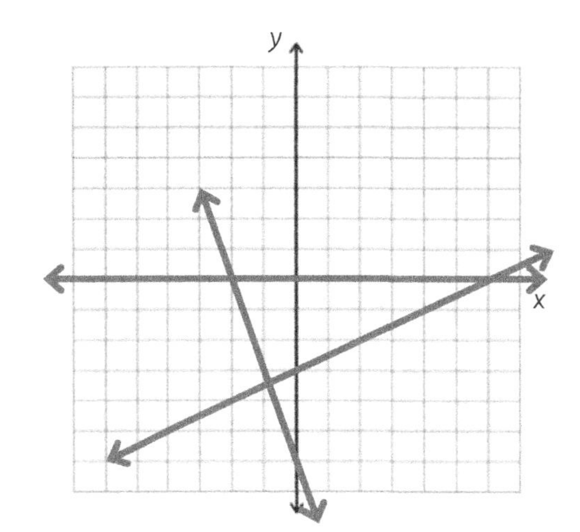

a. Estimate the area of the triangle by counting the squares inside the triangle.

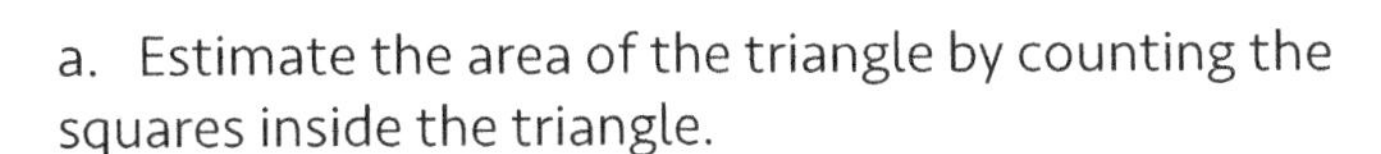

b. Calculate the exact area of the triangle enclosed by the three lines. Express your answer as a mixed number.

23. Consider the triangle formed by the three intersecting lines. Assume that points that look like lattice points are lattice points. Coordinates of a lattice point are integers.

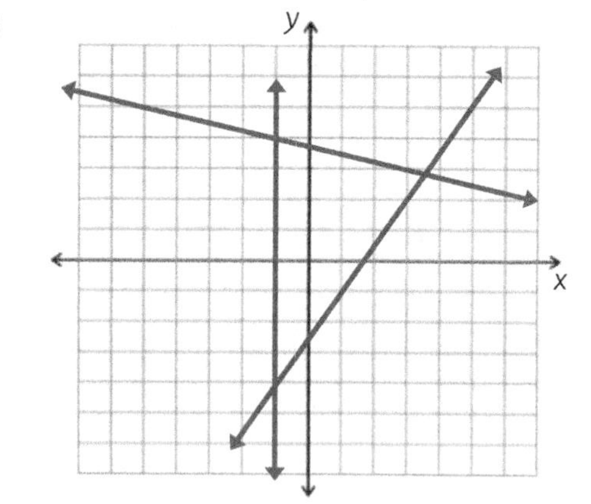

a. Estimate the area of the triangle by counting the squares inside the triangle.

b. ★Calculate the exact area of the triangle enclosed by the three lines. Express your answer as a mixed number.

24. Which triangle has a larger area?

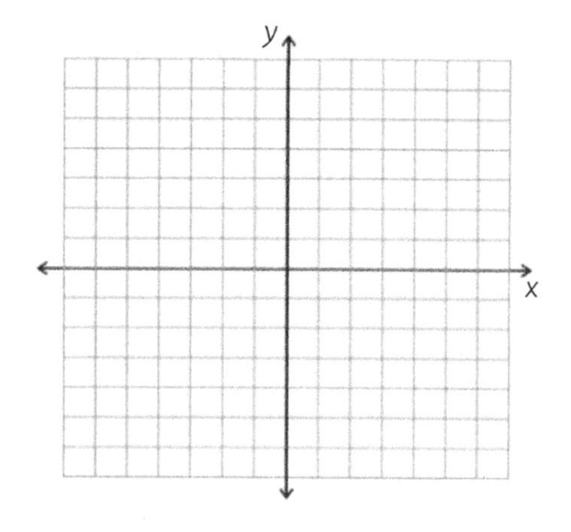

Option #1: The triangle formed by the x-axis and the equations $y=-\frac{5}{2}x-3$ and $y=\frac{1}{2}x+3$.

Option #2: The triangle formed by the y-axis and the equations $y=-\frac{5}{2}x-3$ and $y=\frac{1}{2}x+3$.

25. ★Given the system of equations below, determine the value of the quotient $\frac{x}{y}$.

$21=3y-x$
$3y=-2x-15$

Section 2

SCENARIOS INVOLVING LINEAR SYSTEMS

26. Write the Slope-Intercept Form for a linear equation.

27. Write the Standard Form for a linear equation.

28. Laptop A costs $950 and loses $60 in resale value every year, while Laptop B costs $1100 and loses $100 in resale value every year.

    a. Would you model this scenario with equations in Slope-Intercept Form, equations in Standard Form, or equations in both of these forms?

    b. Which laptop will have a higher resale value after 3 years?

    c. After how many years will the resale value of Laptop A be the same as the resale value of Laptop B?

29. At the end of a busy day at the county fair, an accountant records the earnings. The fair sells tickets for children at $7.50 per ticket and tickets for adults at $10.50 per ticket. On this particular day, 680 people came to the fair and the earnings for the day totaled $6,195.00.

    a. Would you model this scenario with equations in Slope-Intercept Form, equations in Standard Form, or equations in both of these forms?

    b. How many tickets for children were purchased?

30. You find out that there are people in your community who need blankets and toothbrushes so you decide to help them. One week, you spend \$186 and you buy 6 blankets and 24 toothbrushes. The next week, you buy 13 blankets and 6 toothbrushes and spend \$288. How much do the toothbrushes cost and how much do you pay for each blanket?

31. Doug's Rugs charges \$18 per square yard to install new carpeting, with a limited time only \$50 fee for removing old carpeting. Moore's Floors only charges \$16 per square yard to install new carpeting, but they have a \$225 fee for removing old carpeting. How many square yards would you need to install to make Moore's Floors the cheaper company to hire for the job?

32. ★In the previous scenario, how much would you pay to replace the carpeting in a room measuring 6 by 14 yards?

33. The Stellar Cellular Co. charges \$80 per month with an extra charge of \$0.20 per text message after you send 500 text messages. The Cell Allure Phone Co. charges \$70 per month with an extra charge of \$0.24 per text message after you send 500 messages.

    a. If you send 600 text messages one month, which company's plan would you want?

    b. How many text messages would you need to send in one month in order for both plans to have the same total cost?

34. ★After a family photography session, you look through the possible ordering options and notice many combinations. Four combinations are listed below. How much you would pay if you selected a combination that consisted of one 8x10 and one 3x5 photo?

Price	Size and Quantity Combination
$28.00	four 8×10, six 5×7
$31.00	eight 8×10, two 5×7
$22.00	four 3×5, twelve 2×3
$16.00	two 3×5, ten 2×3

35. A pipe is turned on and begins filling a tank with water. After two and a half minutes, another pipe is turned on and begins filling another tank with water. Both pipes are turned off when the tanks contain the same amount of water.

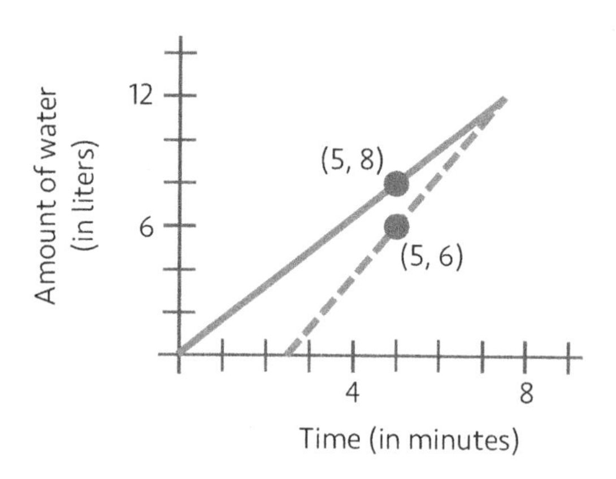

a. What is the flow rate of each pipe, measured in liters per minute?

b. For how many minutes was the second pipe turned on?

c. How much water is in each tank when the pipes are turned off?

36. 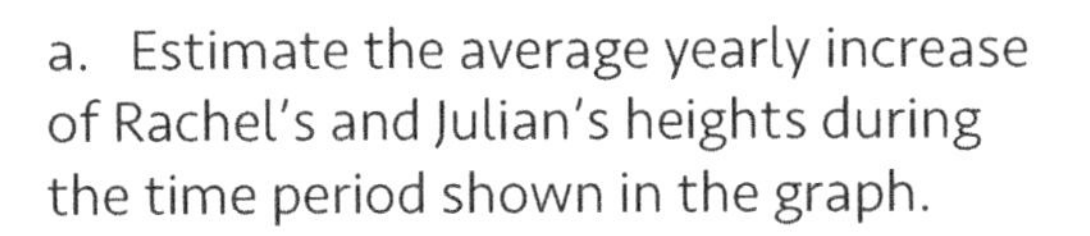
Julian is younger than his older sister Rachel. The chart shows the changes in their heights from 2009 and 2017.

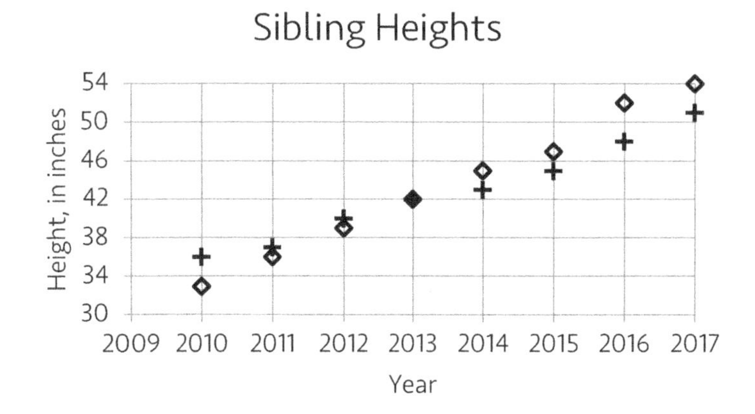

a. Estimate the average yearly increase of Rachel's and Julian's heights during the time period shown in the graph.

b. Draw estimated trend lines for the two sets of data.

c. Find the slopes of both of those lines, rounded to the nearest tenth. What is the significance of finding the slopes of these lines?

37. The previous scenario involves a system of linear equations. A linear system cannot have more than one solution because two distinct lines can only intersect at one point.

a. Describe the graph of a linear system that has no solution.

b. Describe the graph of a linear system that has infinite solutions.

38. ★In order to take care of your new puppy, you set aside money every month to pay for food, chew toys, and training classes. The amount of money that you need to pay for training classes is 120% more than the amount of money you need for food and chew toys, combined. You spend a total of $112 each month on these three expenses.

a. How much do you spend on training classes each month?

b. What percent of your total expenses pay for the training classes?

39. The length of a rectangle is 3 more than twice its width. The perimeter of the rectangle is 48 inches. What is the area of the rectangle?

40. You have \$11,000 in two investment accounts. During that year, you earn 5% interest on the money in one account and 8% interest on the money in the other account. If you earn a total of \$664 in interest that year, how much money did you have in each account at the end of that year?

41. Nyla gets a financial prize after she wins a contest. She puts some of the money in a mutual fund and she puts the rest in bonds. The equations below show the relationship between the amount of money she invested and the amount of interest she earned after one year.

Equation 1: $m + b = 7{,}000$
Equation 2: $0.09m + 0.035b = 509$

a. How much money did Nyla receive for winning the contest?

b. What was the combined amount that Nyla earned in interest after one year?

c. By what percent did the amount of money she invested in bonds increase after one year?

42. In the previous scenario, how much money did Nyla invest in the mutual fund?

Section 3

SYSTEMS OF LINEAR INEQUALITIES

43. Write the inequality that has the solution set shown in each graph below.

a.

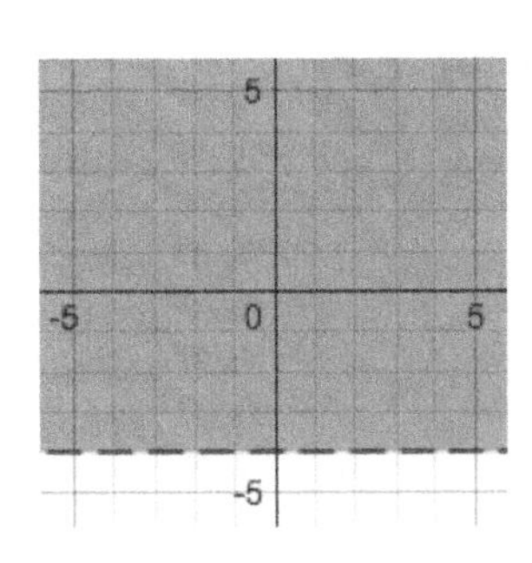

b.

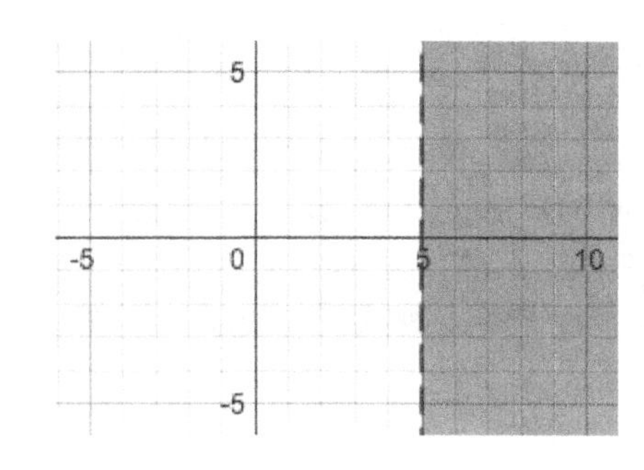

44. Graph each inequality.

a. $x<4$

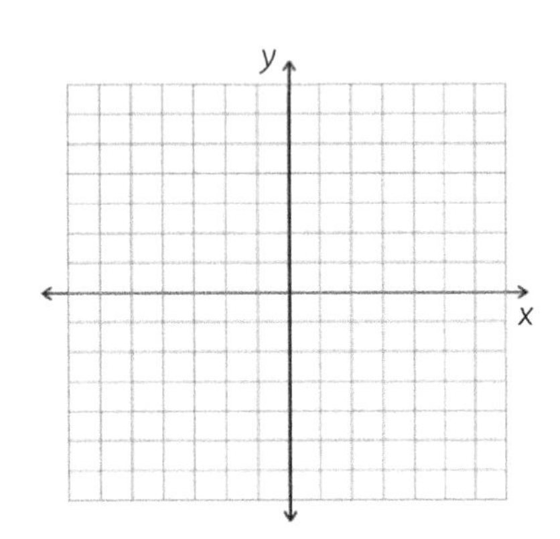

b. $y \leq -2$

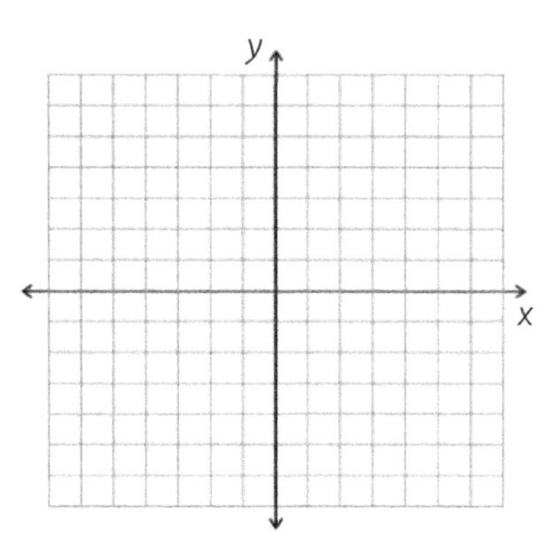

45. Circle the inequalities that would have a shaded region that includes the boundary line.

a. $y \geq 2x+7$ b. $x>5$ c. $y-x<1$ d. $-3x+2y \leq 6$

46. Graph each inequality.

a. $y \geq -2x+4$

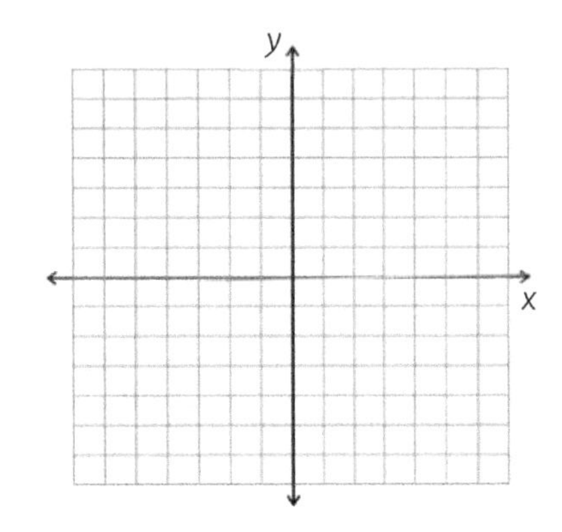

b. $x-4y \geq 12$

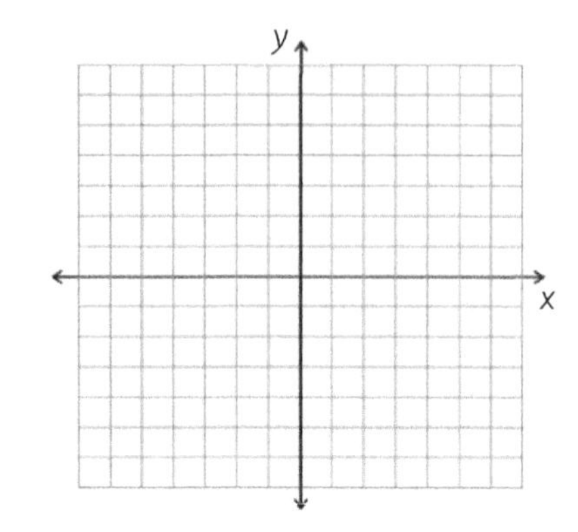

47. Write the inequality that has the solution set shown in each graph below.

a.

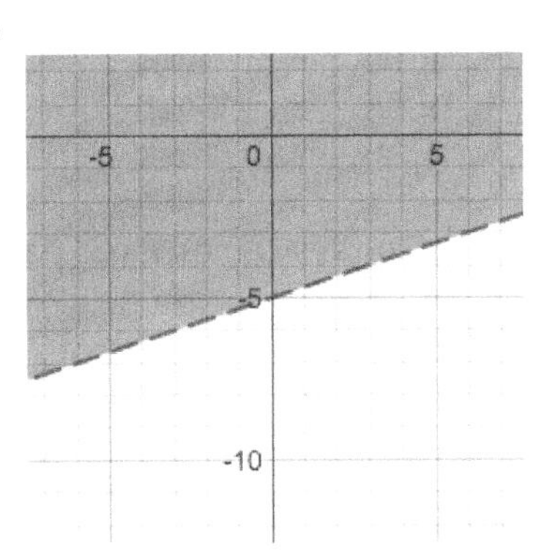

b.

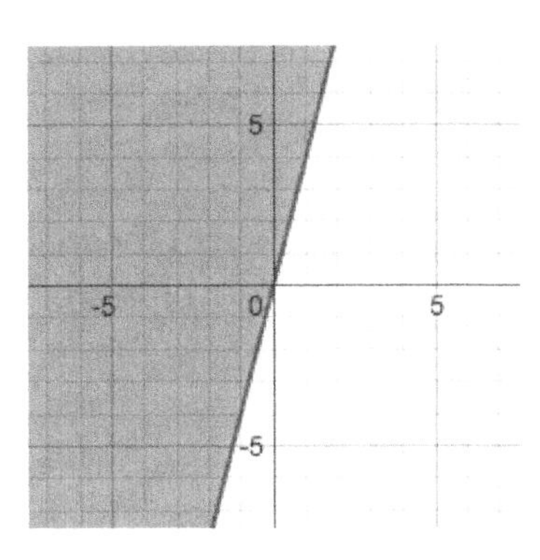

c.

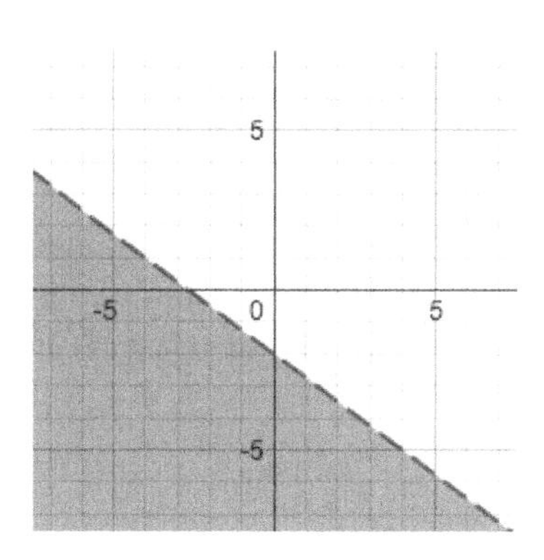

48. Now consider an inequality that has 2 restrictions. In the graph to the right, plot as many points as you can that satisfy both of the conditions below.

The x-value is less than 2.

The y-value is less than −2.

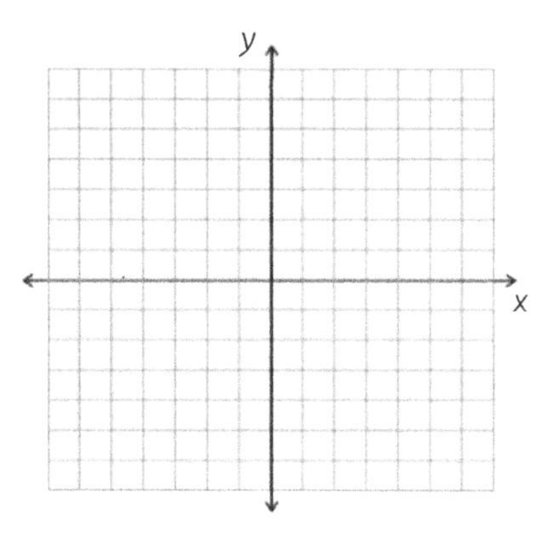

49. The lines in the graph shown divide the plane into 4 regions. Shade in the region that satisfies the 2 inequalities below.

$x \geq -1$

$y \leq 4$

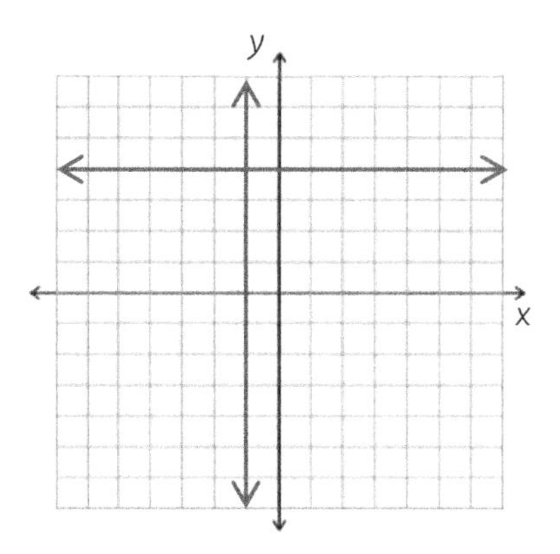

50. The lines in the graph shown divide the plane into 4 regions. Shade in the region that satisfies the 2 inequalities below.

$y < -\frac{1}{2}x + 3$

$y > \frac{1}{2}x - 1$

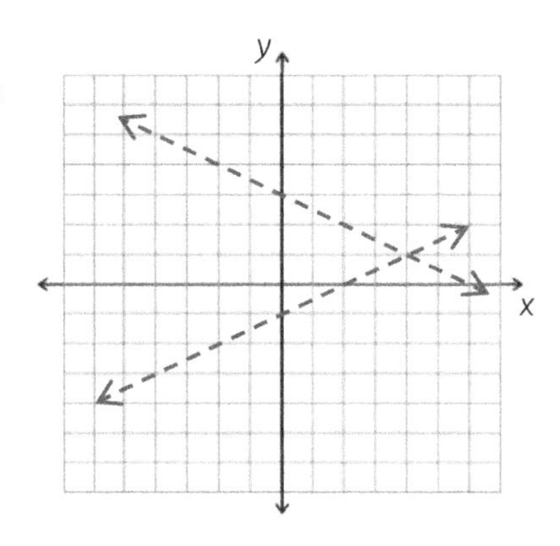

51. How would you change your graph in the previous scenario to display the solution region if the inequalities were changed to become $y \le -\frac{1}{2}x+3$ and $y \ge \frac{1}{2}x-1$?

When you graph a linear inequality, the darkened (or shaded) region displays the ordered pairs that make the inequality true. When you graph two linear inequalities, the shaded region displays the ordered pairs that make both inequalities true.

52. Display the solution of the system of inequalities. This is also known as graphing the system.

$$y > -\frac{2}{3}x-1$$
$$y \le \frac{3}{2}x-3$$

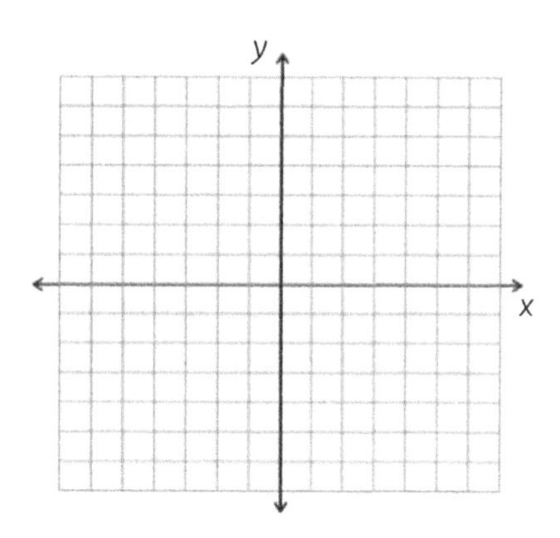

53. Graph the system of inequalities.

$$4x+y>4$$
$$4x-y \ge 4$$

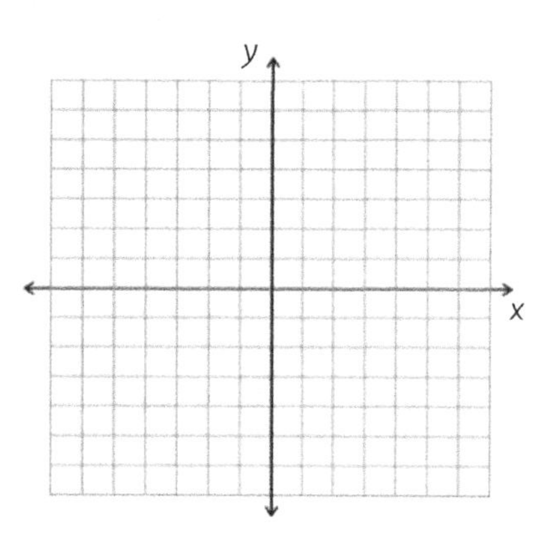

54. Graph each system of inequalities.

a. $y>3x-5$
$y-3x<2$

b. $y \ge \frac{3}{4}x+3$
$3x-4y>8$

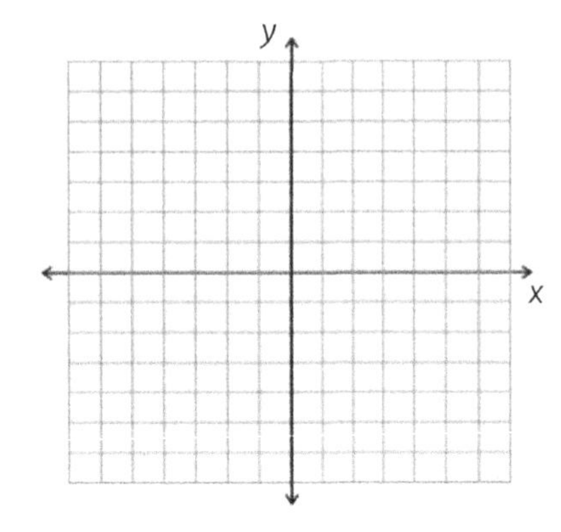

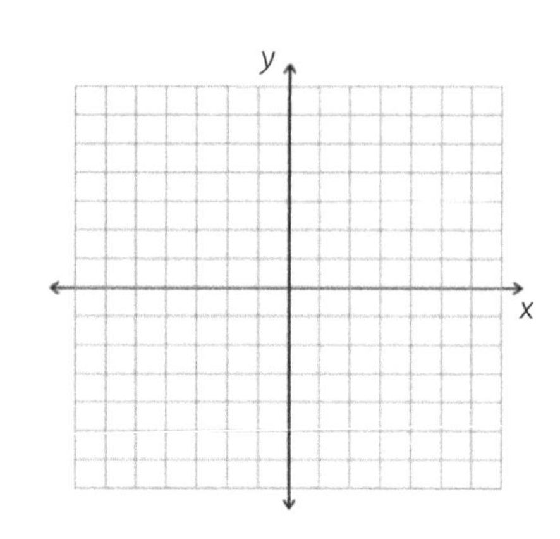

55. Graph the system of inequalities using only the graph shown. Do not change the numbering on the axes.

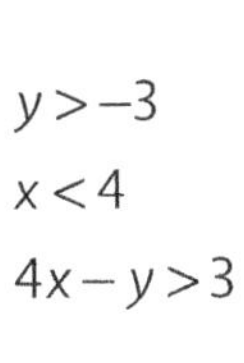

$y > -3$

$x < 4$

$4x - y > 3$

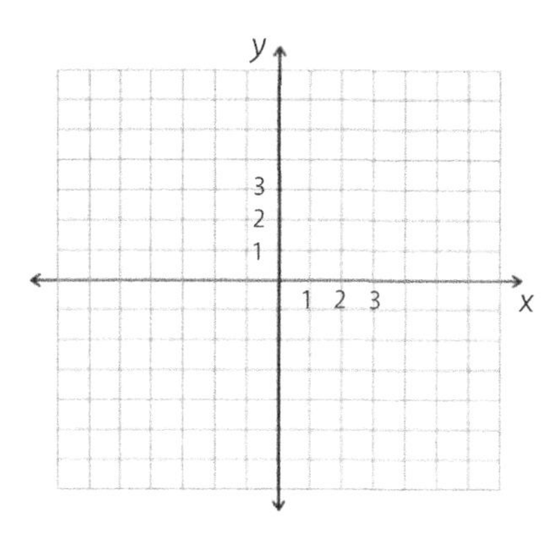

56. Calculate the area of the triangular shaded region in the previous scenario.

57. Write a system of inequalities whose solution set is shown by the shaded region.

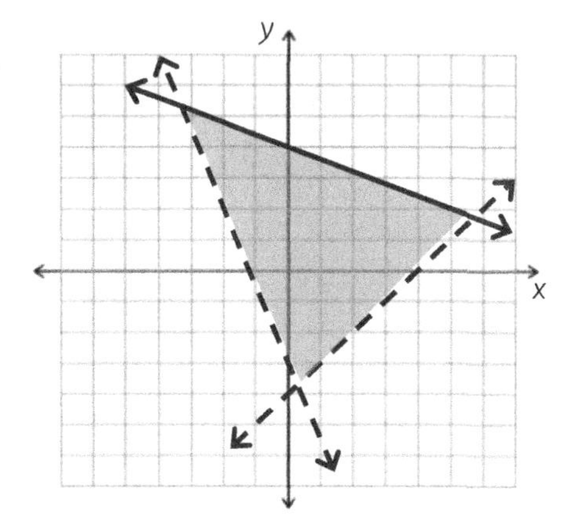

58. The equations of four lines are written below and their graphs are shown to the right.

$4x = -3y - 12, \quad 3x + 4y = 12, \quad 3y - 4x = 9, \quad 4y = 3x + 28$

Determine the coordinates of H.

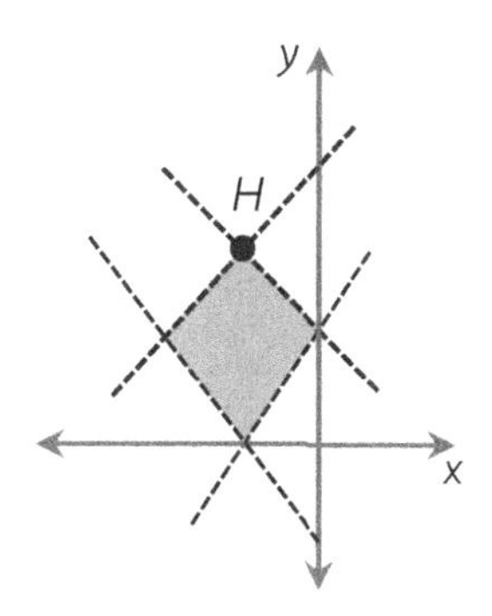

59. Write a system of inequalities for the shaded region in the previous scenario.

60. Calculate the area of the region defined by the following inequalities.

$x - y > -3$

$y > -2$

$y < -\frac{3}{2}x + 1$

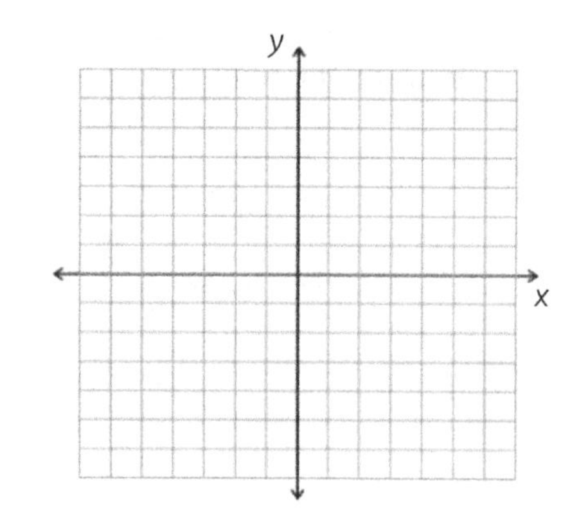

61. With your new gift card, you can either buy songs for \$1 each or movies for \$3 each. You want to buy at least 6 items but the value of your gift card is only \$10. How many songs and movies can you buy?

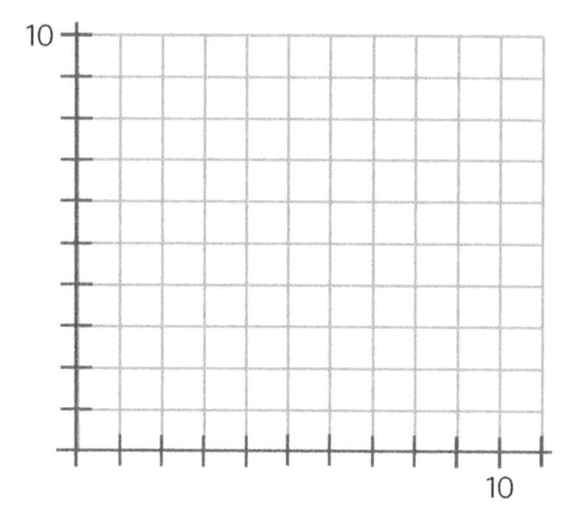

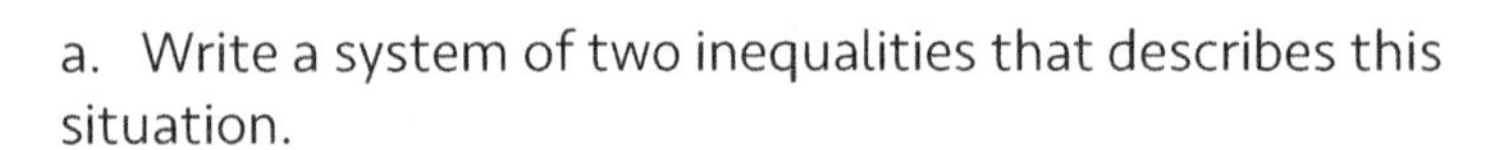

a. Write a system of two inequalities that describes this situation.

b. Graph the system to show all possible solutions.

c. Write two possible solutions to the scenario.

★d. If one possible combination is 1 movie and 6 songs, how many different combinations are possible?

62. Create a system of inequalities that satisfies the given conditions. Graph the system to check your solution.

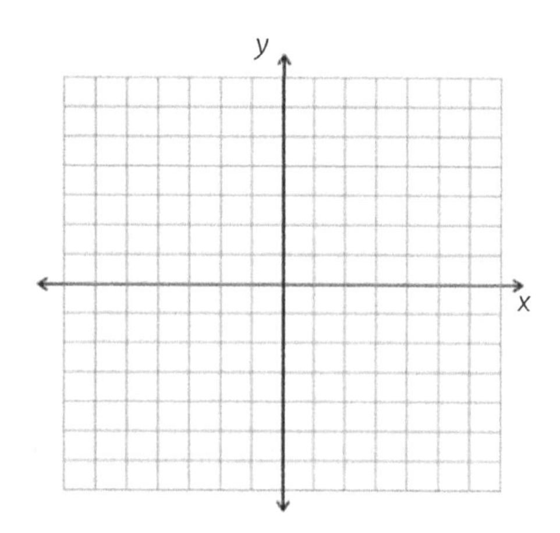

Conditions:
1. The shaded region forms a right triangle.

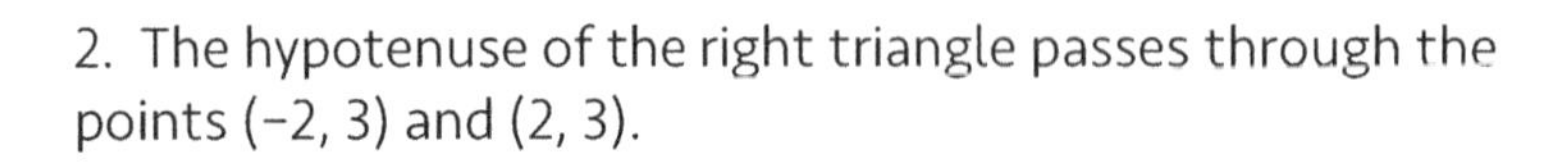

2. The hypotenuse of the right triangle passes through the points (−2, 3) and (2, 3).

3. One vertex of the triangle is located at (1, −3) and another vertex of the triangle is located at (−5, 3).

Section 4
NONLINEAR SYSTEMS

In previous lessons, you have learned about linear systems. In the next group of scenarios, you will learn that systems can involve equations that are nonlinear. These systems may contain more than one solution but you can find these solutions using methods that you already know.

63. Simplify each expression below.

a. $(x-11)^2$ b. $(7y-3)^2$ c. $(t^2+1)^2$

64. If $y=x-2$, what is the value of x^2?

65. If $x=-6$, what is the value of $-\frac{1}{2}x^2-x+10$?

66. Solve the equation $2x^3-8x=0$ by factoring.

67. Consider the system shown to the right. Identify the approximate solutions of this system.

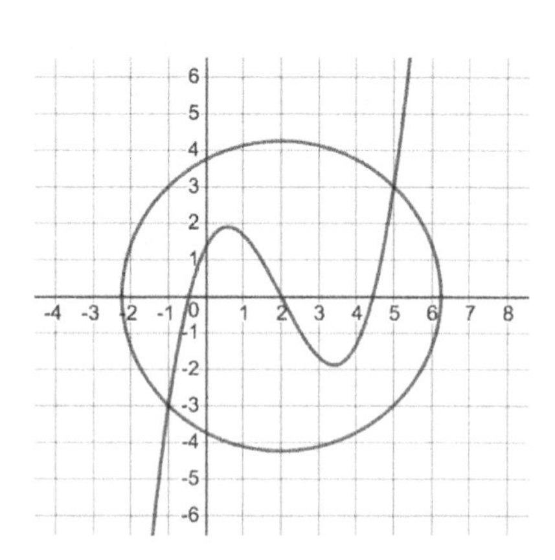

68. A nonlinear system of equations is shown in the graph. How many solutions does this system have?

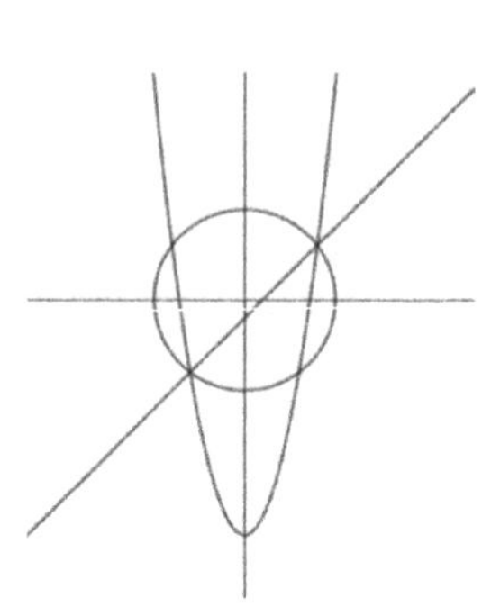

69. A system of two equations is shown below. A portion of their graphs is also shown. Without using the graphs, how can you find the intersection point of these two parabolas?

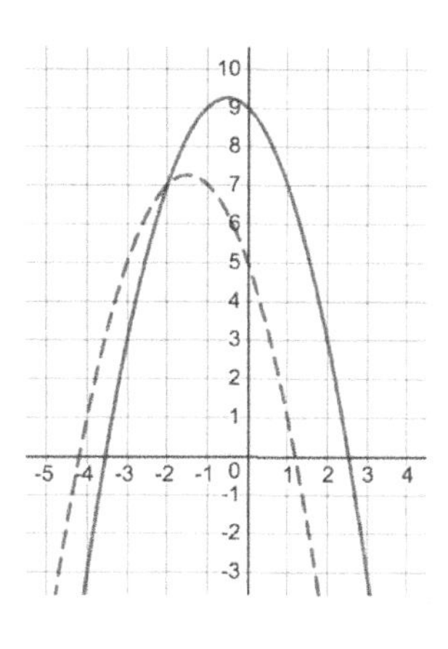

$$y=-x^2-x+9$$
$$y=-x^2-3x+5$$

70. Find the intersection point of the parabolas in the previous scenario using the method you described in the previous scenario.

71. A system of two equations is shown below. Find the intersection point(s) of these two parabolas.

$$y=x^2-2x+3$$
$$y=2x^2-7x-3$$

72. A system of two equations is shown below. A portion of their graphs is also shown in the Cartesian plane to the right. Use algebra to find the exact the solutions of this system. You must verify your result with work. (Do not use the grid lines to estimate the solutions. The lines are intentionally confusing.)

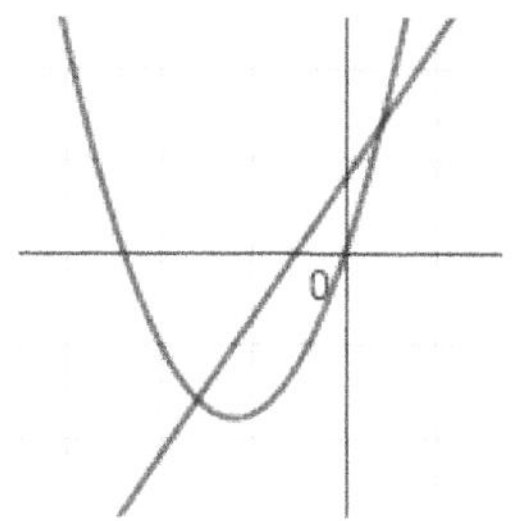

$$y=x^2+3x$$
$$y=\frac{3}{2}x+1$$

73. Where do the two functions intersect? Determine the exact coordinates. Verify the accuracy of your results by displaying the functions on a graphing application.

$$y=-2x^2+2x+3$$
$$y=-x+4$$

74. ★Where do the two functions intersect? Determine the exact coordinates. Verify the accuracy of your results by displaying the functions on a graphing application.

$$f(x)=-x^3+4x^2-2x+3$$
$$g(x)=x^2-2x+3$$

75. Solve the system shown below and verify your results by using a graphing application.

$$\begin{cases} x^2+y^2=100 \\ y=-\frac{3}{4}x \end{cases}$$

76. Given the system of equations below, what is the value of x^2?

$$x^2 + y^2 = 120$$
$$y = 2x$$

77. Draw a system that consists of a line and a circle with the following characteristic.

a. no solution

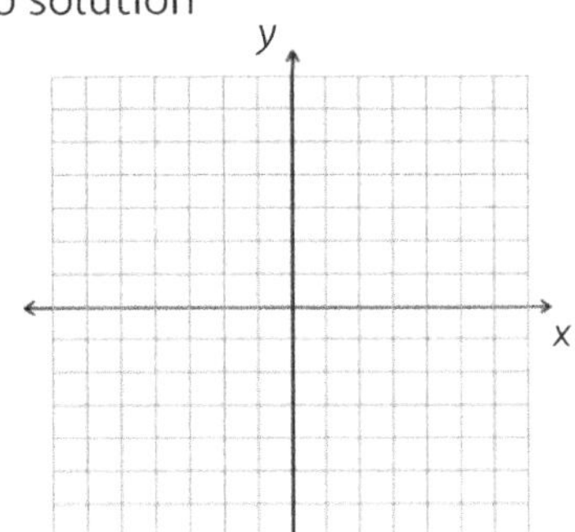

b. one solution

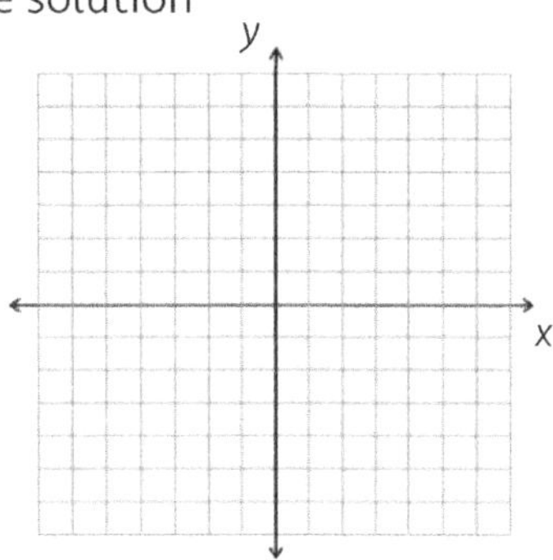

78. Draw a system that consists of a line and a circle with the following characteristic.

a. two solutions

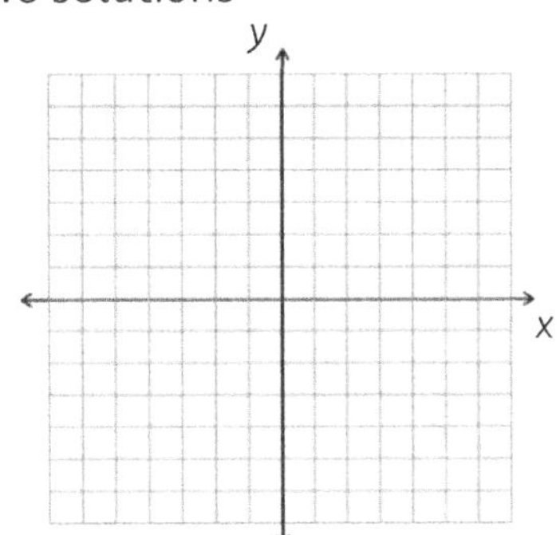

b. three solutions

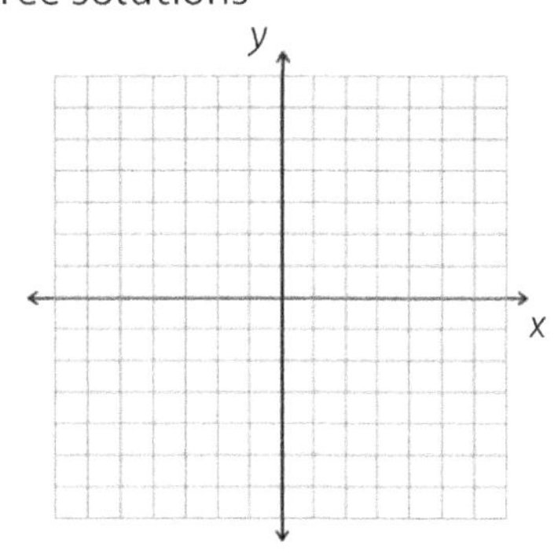

79. Solve the system shown below and verify your results by using a graphing application.

$$\begin{cases} y = -\frac{1}{2}x + 1 \\ x^2 + y^2 = 25 \end{cases}$$

80. Solve the system shown below and verify your results by using a graphing application.

$$\begin{cases} y^2 = 100 - x^2 \\ x^2 + y^2 = 64 \end{cases}$$

81. Why does the previous system have no solution?

82. ★Solve the system shown below and verify your results by using a graphing application.

$$\begin{cases} x^2 + y^2 = 100 \\ y = \frac{3}{2}x - 6 \end{cases}$$

83. A company makes and sells cups. There is a cost associated with producing the cups, but the company overcomes these costs by selling the cups at a high enough price. The graphs for the revenue, $R(n)$, and costs, $C(n)$, are shown, where n represents the number of cups that have been produced and sold in a given month.

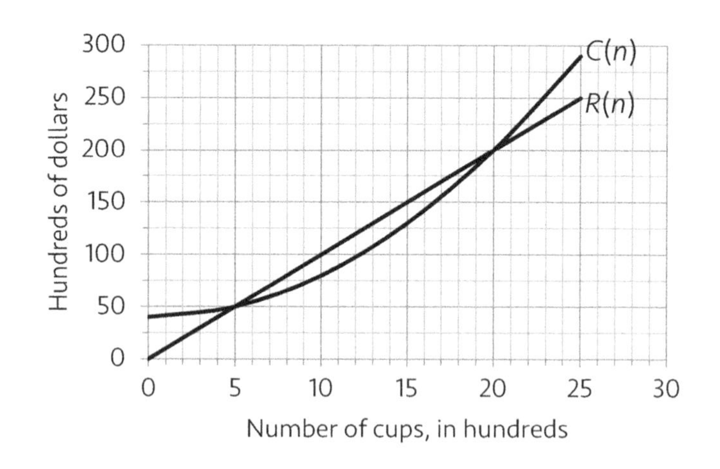

a. What is $R(15)$? What does it mean?

b. Estimate $C(23)$. What does it mean?

84. A company makes a profit when its revenue is more than its costs. In the previous scenario, how many cups need to be produced and sold for the company to make a profit for the month?

Section 5

SYSTEMS WITH 3 VARIABLES

Two approaches that can be used to solve a system of equations with two unknowns are the Substitution Method and the Elimination Method. In the Substitution Method, you isolate one variable and make a substitution into another equation to replace that variable with an equivalent expression. In the Elimination Method, you combine two equations through addition or subtraction in such a way that one of the variables disappears, leaving behind a one-variable equation.

85. In each equation shown, isolate the requested variable.

a. Solve for x.

$x-7y-3z=4$

b. Solve for y.

$6x+y-z=11$

c. Solve for z.

$5x+5y-z=1$

86. Next, make a substitution as requested. Notice that the substitution converts a 3-variable equation into a 2-variable equation.

a. Substitute $x=2y+z+7$ into the equation $3x+y-2z=5$. Simplify the result.

b. Substitute $y=2x-z-5$ into the equation $4x-y-3z=2$. Simplify the result.

87. Now, use what you know about the Elimination Method and apply that to 3-variable equations. Notice that the elimination converts a 3-variable equation into a 2-variable equation.

a. Eliminate x.

$2x+6y+z=10$
$2x+3y+2z=1$

b. Eliminate y.

$3x-4y+2z=-3$
$-x+y+6z=5$

c. Eliminate z.

$-x+5y-4z=7$
$5x-2y+3z=-9$

88. Solve the 3-variable system of equations using the Substitution Method.

$$3x - y - 2z = -1$$
$$5x - 3y + 5z = 26$$
$$-2x + y - z = -7$$

89. It may be helpful to relate this topic to something familiar. When a 2-variable equation is in the form $Ax + By = C$, it looks like a line when graphed. When you solve a system of linear equations, you are essentially trying to find the intersection point of two lines.

a. What geometric shape is formed by a 3-variable equation in the form $Ax + By + Cz = D$?

b. When you solve a 3-variable system, what does the solution represent?

90. Solve the 3-variable system of equations using the Substitution Method.

$$3x + 2y + 4z = 11$$
$$2x - y + 3z = 4$$
$$5x - 3y + 5z = -1$$

91. Solve the 3-variable system of equations using the Elimination Method.

$$-3x+y+2z=1$$
$$4x-2y+2z=14$$
$$-5x+3y-5z=-26$$

92. ★Solve the 3-variable system of equations using the Elimination Method.

$$-2x+y+6z=1$$
$$3x+2y+5z=16$$
$$7x+3y-4z=11$$

93. Solve the 3-variable system of equations using any combination of methods.

$$2x-4y-z=-18$$
$$-6x-3y+2z=2$$
$$4x+y-6z=-37$$

94. A system of three equations is shown below. A portion of their graphs is also shown in the Cartesian plane to the right.

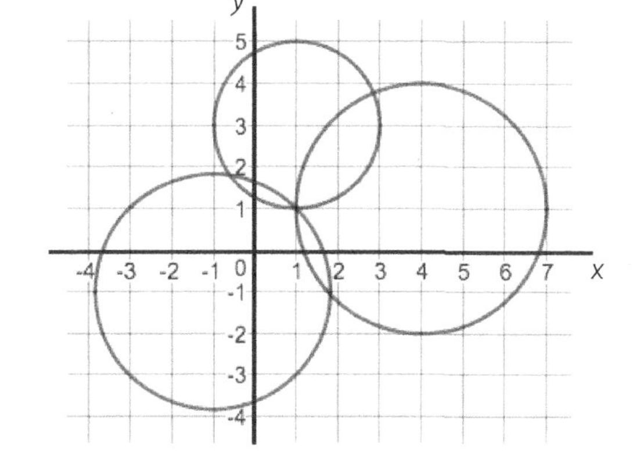

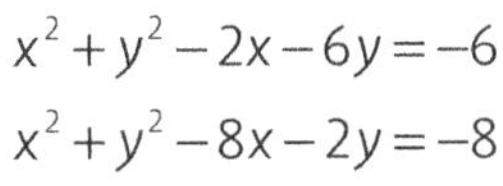

$x^2+y^2-2x-6y=-6$

$x^2+y^2-8x-2y=-8$

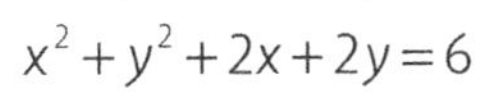

$x^2+y^2+2x+2y=6$

a. How many solutions does this system have?

b. Estimate the solution of this system.

95. ★Use what you have learned about solving systems of equations to verify that the three circles in the previous scenario intersect at (1, 1).

96. To analyze how pricing affects a person's purchasing decisions, a company puts water in 3 types of bottles and sells it in a busy city on a hot summer day. The water in a blue bottle sells for \$2, while the green bottle is \$3 and the yellow bottle is \$3.50. At the end of the day, they had sold 19,200 bottles and earned \$56,400 in revenue. The number of green bottles purchased was three times larger than the number of blue bottles purchased.

a. How many yellow bottles were purchased?

★b. What percent of the total revenue was due to green bottle purchases?

Section 6

FINDING THE EQUATION FOR A PARABOLA, GIVEN 3 POINTS

97. Consider three points that are part of a parabola: (−3, 3), (−2, 6), (−1, 7). Begin the process of finding the quadratic function, $y = Ax^2 + Bx + C$, that contains these points. Plug in each of these three points into the structure $y = Ax^2 + Bx + C$ to create three equations that contain *A*, *B*, and *C*. Do not solve these equations. Just write them.

98. Plot the previous three points in the graph shown.

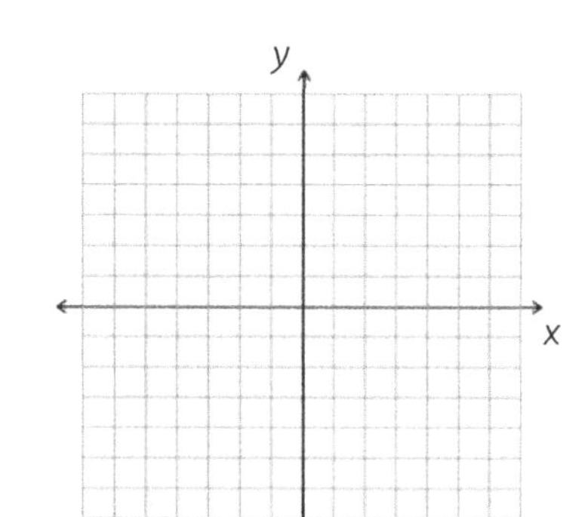

99. Find the equation for the parabola that passes through the following points.

$$(-3, 3), (-2, 6), (-1, 7)$$

100. Complete the graph of the previous parabola by plotting at least 4 more points and then draw the parabola that passes through those points.

101. Find the equation for the parabola that passes through the following points.

$$(-4, -9), (-2, 5), (1, -4)$$

102. Consider the table below.

x	−1	0	1	2	3
y	−4	−3	0	5	

a. Identify a pattern in the table above and use the pattern to find the missing term.

b. The ordered pairs are points from a parabola. Find the equation of this parabola.

103. Determine the equation of the parabola shown. Write the equation in Standard Form.

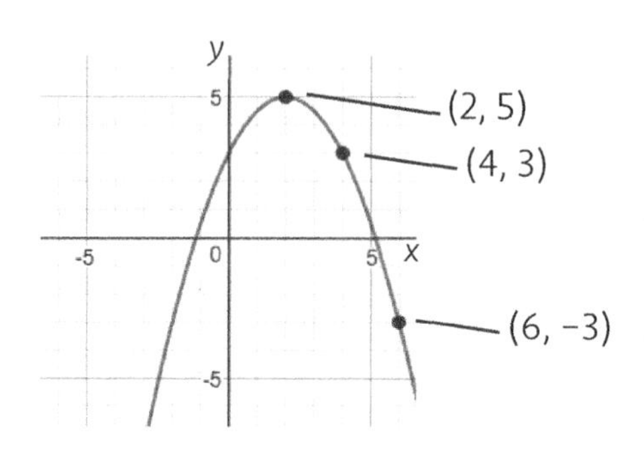

104. ★Gabby puts on a harness, attaches herself to a bungee cord, and then jumps off of a bridge that runs over part of a lake far below. After she has been falling for one-half of a second, she is 140 feet above the water below. After falling for 1 second, she is 128 feet above the water. After 2 seconds, her height is 80 feet.

a. If she forgot to attach the other end of the bungee cord to the bridge, for how many seconds will she fall before she splashes into the lake? (Fortunately, she was not injured when she hit the water.)

b. How high above the ground was Gabby at the moment she jumped?

105. The equation for a parabola is shown below.

$$f(x) = -2x^2 + 8x - 10$$

a. Find the coordinates of the vertex of the parabola.

b. Where does the parabola cross the x-axis?

106. ★Part of a parabola is shown in the graph.

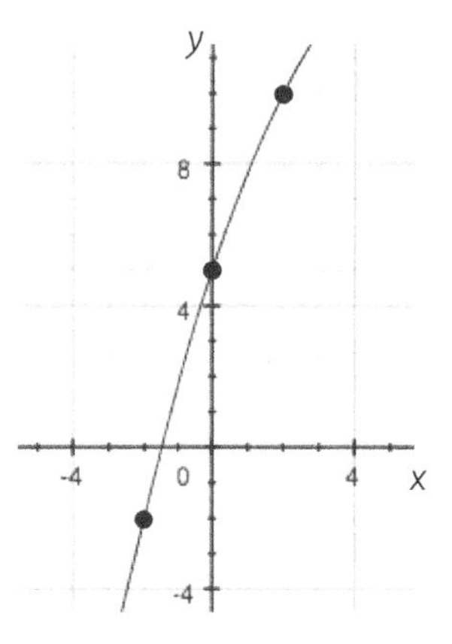

a. Determine the coordinates of the vertex of the parabola.

b. Identify the exact x-intercepts of the parabola.

c. Try to draw more of the parabola. Do not try to be exact. Instead, estimate the shape of the curve as you extend the part that is shown on the graph.

Section 7
CUMULATIVE REVIEW

107. Write out The Quadratic Formula.

108. How can you use a parabola's equation to determine if a parabola will open upward or downward?

109. Determine the x-intercepts of the function $f(x)=-\frac{1}{2}x^2+x-1$.

110. If the vertex of a parabola is located at (−2, 7) and one of the x-intercepts is (3, 0), where is the other x-intercept located?

111. Graph the function $f(x)=x^2-2x-6$. Plot at least 7 points, including the vertex, the x-intercept(s), and the y intercept.

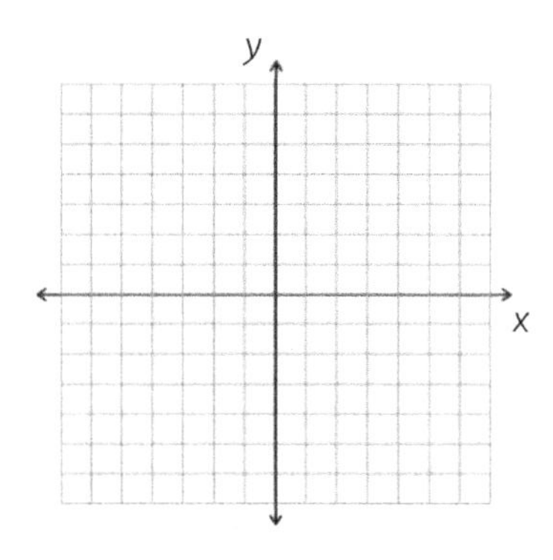

112. Calculate the area of the right triangle formed by the system of inequalities shown.

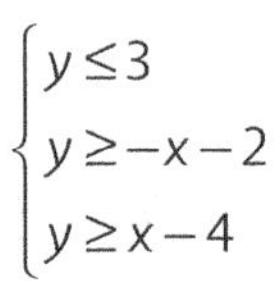

$$\begin{cases} y \le 3 \\ y \ge -x-2 \\ y \ge x-4 \end{cases}$$

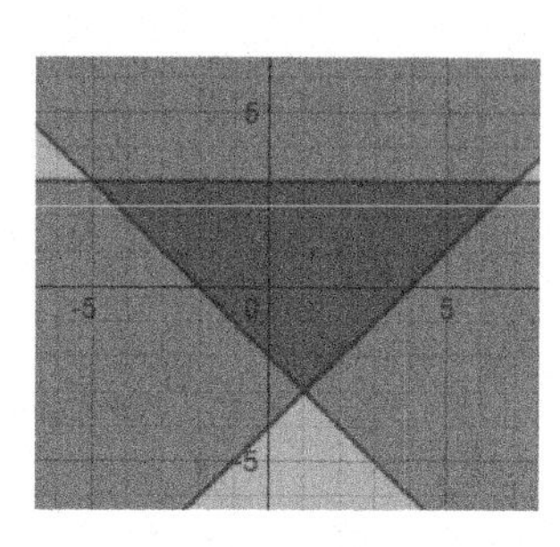

113. ★A cleaning solution is made by filling a bucket with water and then adding vinegar until the bucket contains 4 gallons of liquid. At this point, the solution is 20% vinegar. When the solution is tested, it is too diluted so more vinegar is added until the bucket contains 50% vinegar. How much vinegar was added to the original solution?

114. After the vinegar is added to the original solution in the previous scenario, a sponge is used to clean the floors in a room. If the sponge is dipped into the bucket every 24 seconds, on average, and it takes 32 minutes to empty the entire cleaning solution, how much liquid does the sponge draw out of the bucket each time it is dipped into the solution?

115. Derek's first job is at a car dealership that pays him $1,000 per month in addition to 10% of the money that he earns from his car sales for the month. Sonia's first job is for a clothing company that pays her $1,300 per month plus 5% of her total monthly sales. One month, they compare their results and find out that their sales totals are the same, as well as their paychecks for that month. What was the amount written on their paycheck?

116. You invest $1000 for one year and earn 4% interest on your investment. You also deposit $4000 in a savings account that earns 0.5% interest and leave it in the account for one year.

a. By what percent did the total value of your accounts increase that year?

b. If you had switched the amounts of money that you placed in the two accounts, by what percent would the total value of your accounts have increased?

117. Several years later, you have a total of $8,900 in your two accounts. During that year, you earn 5% interest on your invested money and you earn 1.2% interest on the money in your savings account. If you earn a total of $350 in interest that year, how much money did you place in investments and in savings that year, respectively?

118. The graph of a quadratic function is shown. On the parabola, find the point with the greatest y-value. What is the x-value at this point?

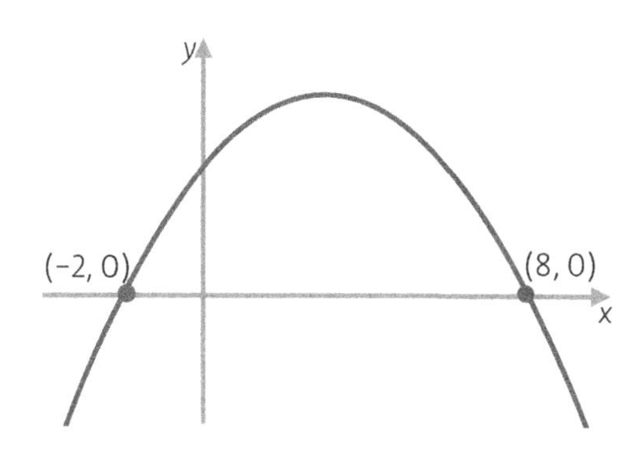

119. Solve the equation.

$$14-7m^2=-49$$

Section 8
ANSWER KEY

1.	(−4, 1)
2.	The variable "y" is isolated in the first equation ($y=-2x+5$) so the "y" in the second equation can be replaced with "$-2x+5$".
3.	(2, 1)
4.	(0, 3)
5.	a. (1, 7) b. (26, 4)
6.	(8, 6)
7.	The "y" terms in both equations have opposite coefficients. Without changing either of them, the two equations can be added to eliminate the y terms.
8.	(4, 1)
9.	a. (2, 1) b. (2, −10)
10.	(1, 2)
11.	(8, 6)
12.	$\left(-3\frac{1}{9}, 1\frac{1}{9}\right)$
13.	a. $1\frac{1}{9}$ units b. $3\frac{1}{9}$ units
14.	Everywhere (they are the same line)
15.	a. $y=-\frac{1}{2}x-\frac{5}{2}$ b. lines are parallel
16.	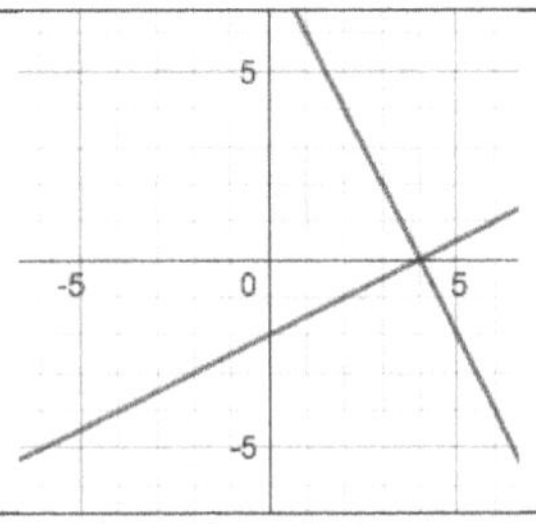 $y=\frac{1}{2}x-2$
17.	Yes, they intersect at (10, 11).
18.	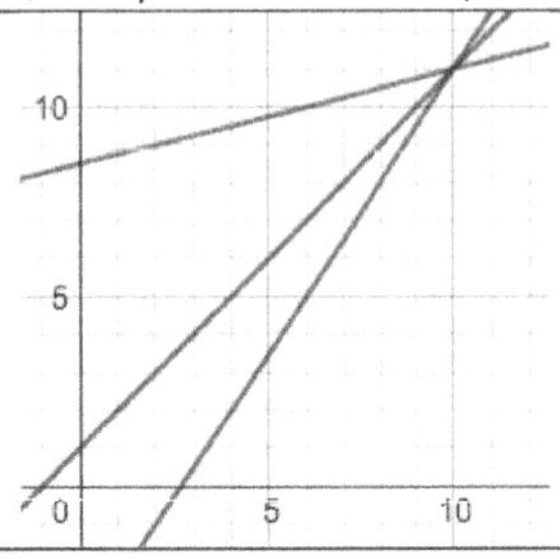
19.	a. $11 \cdot 7 = 77 \text{ in.}^2$ b. $\frac{1}{2}(11 \cdot 7) = 38.5 \text{ cm}^2$
	c. $\frac{11 \cdot 7}{2} = 38.5 \text{ m}^2$
20.	Area $= \frac{1}{2}(9)(6) = 27$ units^2
21.	a. Around 20 units^2 b. Intersection point: (−0.75, 4.5) Area $= \frac{1}{2}(9)(4.5) = 20.25$ units^2
22.	a. Around 13 or 14 units^2 b. Intersection point: $\left(-\frac{6}{7}, -3\frac{3}{7}\right)$ Area $= \frac{1}{2}(8)\left(\frac{24}{7}\right) = 13\frac{5}{7}$ units^2
23.	a. Around 18 or 19 units^2 b. Intersection point: $\left(3\frac{4}{7}, 2\frac{6}{7}\right)$ Area $= \frac{1}{2}(8)\left(4\frac{4}{7}\right) = 18\frac{2}{7}$ units^2
24.	Option #2 (6 units^2 vs. 5 units^2) Both triangles have the same height, but the #2 triangle has a base that is 1 unit longer.
25.	−4; $x = -12$ and $y = 3$
26.	$y = mx + b$
27.	$Ax + By = C$
28.	a. Slope-Intercept Form Laptop A: $V = 950 - 60y$ Laptop B: $V = 1100 - 100y$ b. Laptop B c. 3 years, 9 months Solve: $950 - 60y = 1100 - 100y$; $y = 3.75$
29.	a. Standard Form Equation 1: $a + c = 680$ Equation 2: $10.5a + 7.5c = 6195$ b. 315 Tickets for children
30.	Equation 1: $6b + 24t = 186$ Equation 2: $13b + 6t = 288$ 1 blanket = \$21 1 toothbrush = \$2.50
31.	More than 87.5 square yards Solve: $18y + 50 > 16y + 225$

32.	\$1,562, because you would hire Doug's Rugs. They would be the cheaper company for a floor with an area of 84 square yds.
33.	a. Cell Allure b. 750
34.	Total: \$5 8x10=\$3.25; 3x5=\$1.75
35.	a. 1.6 liters/min and 2.4 liters/min b. 5 min c. 12 liters Equations: $y=\frac{8}{5}x$ and $y=\frac{12}{5}x-6$
36.	a. Rachel: ≈2 in/yr Julian: ≈3 in/yr c. Sister: approx. 2.2; Brother: approx. 3 the slopes are the average increase in the height each year
37.	a. draw parallel lines b. draw the same line twice
38.	a. \$77 (Hint: Let food and chew toys be a single variable, such as C, for combined) Equation 1: $C + T = 112$ Equation 2: $T = C + 1.20C \rightarrow T = 2.2C$ b. 68.75%
39.	Equation 1: $L = 2W + 3$ Equation 2: $2L + 2W = 48$ W = 7 in. L = 17 in. Area = $7\times17 = 119$ in.^{2}
40.	Equation 1: $A + B = 11{,}000$ Equation 2: $0.05A + 0.08B = 664$ A = \$7,200 B = \$3,800
41.	a. \$7,000 b. \$509 c. 3.5%
42.	In the system of equations, solve for m. m = \$4,800
43.	a. $y>-4$ b. $x>5$
44.	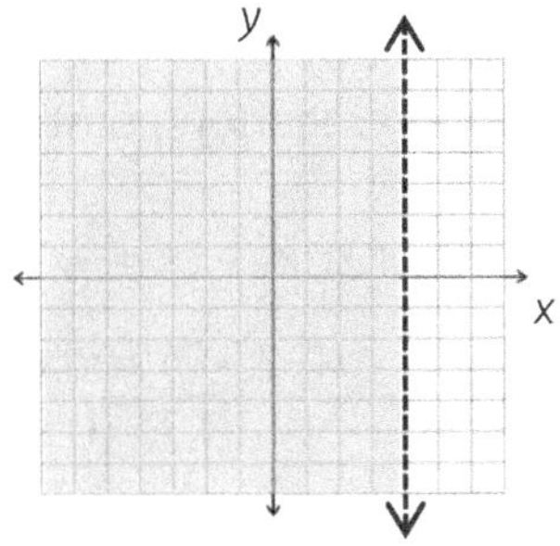 a. 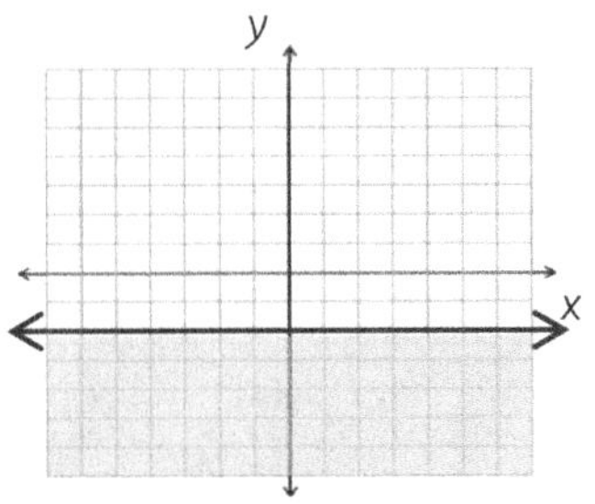 b.
45.	Circle a. and d.

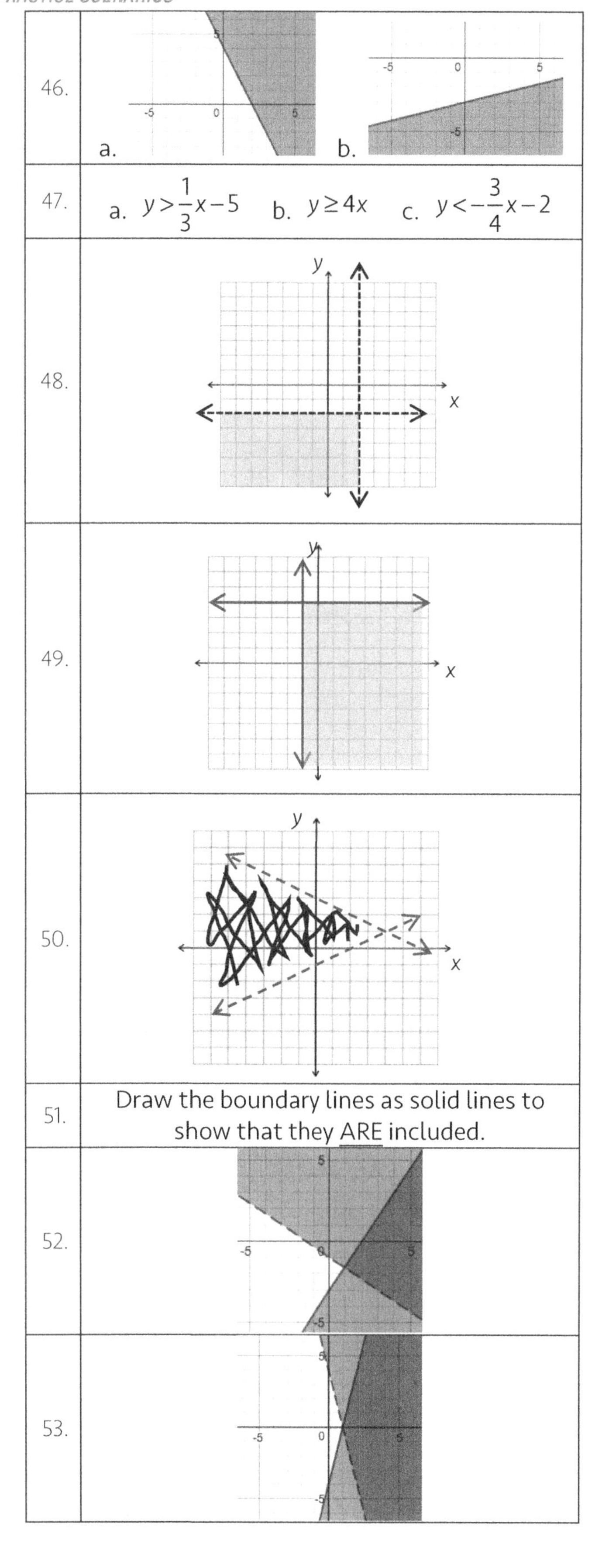

46.	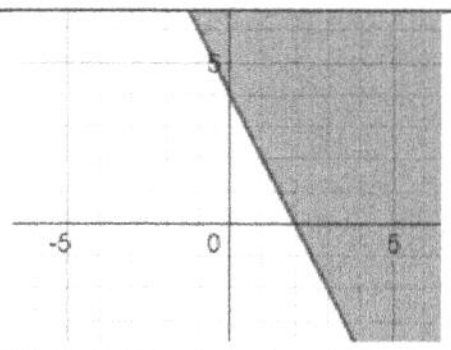 a. 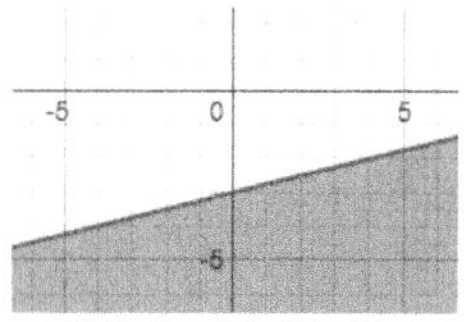 b.
47.	a. $y>\frac{1}{3}x-5$ b. $y\geq 4x$ c. $y<-\frac{3}{4}x-2$
48.	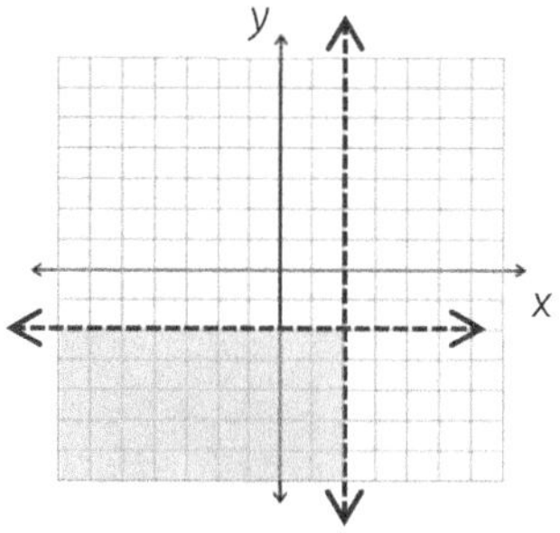
49.	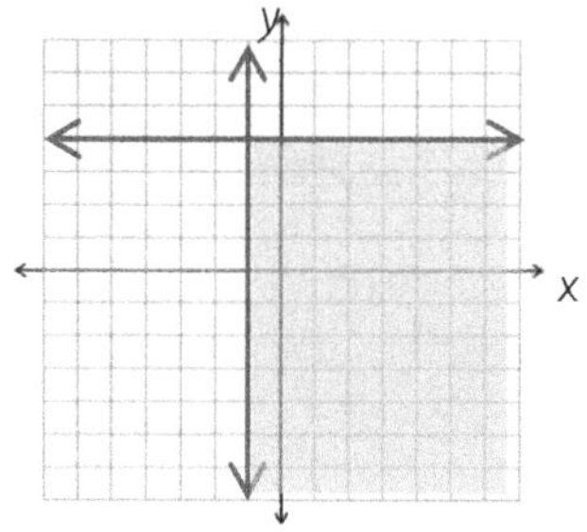
50.	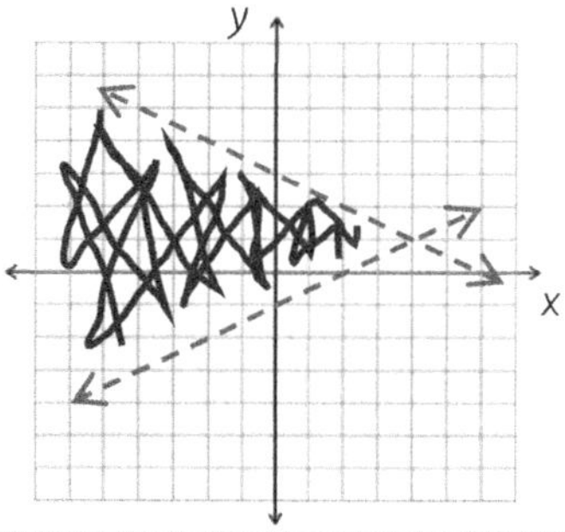
51.	Draw the boundary lines as solid lines to show that they ARE included.
52.	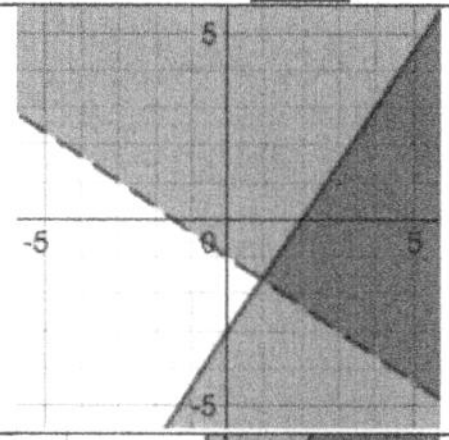
53.	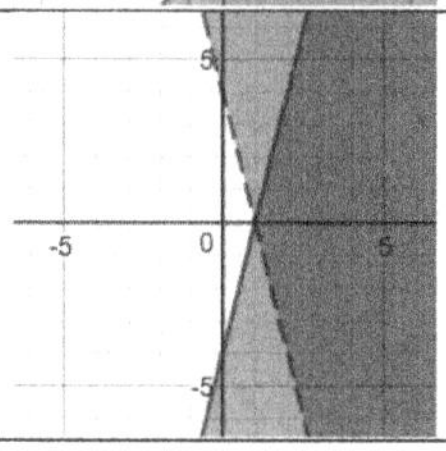

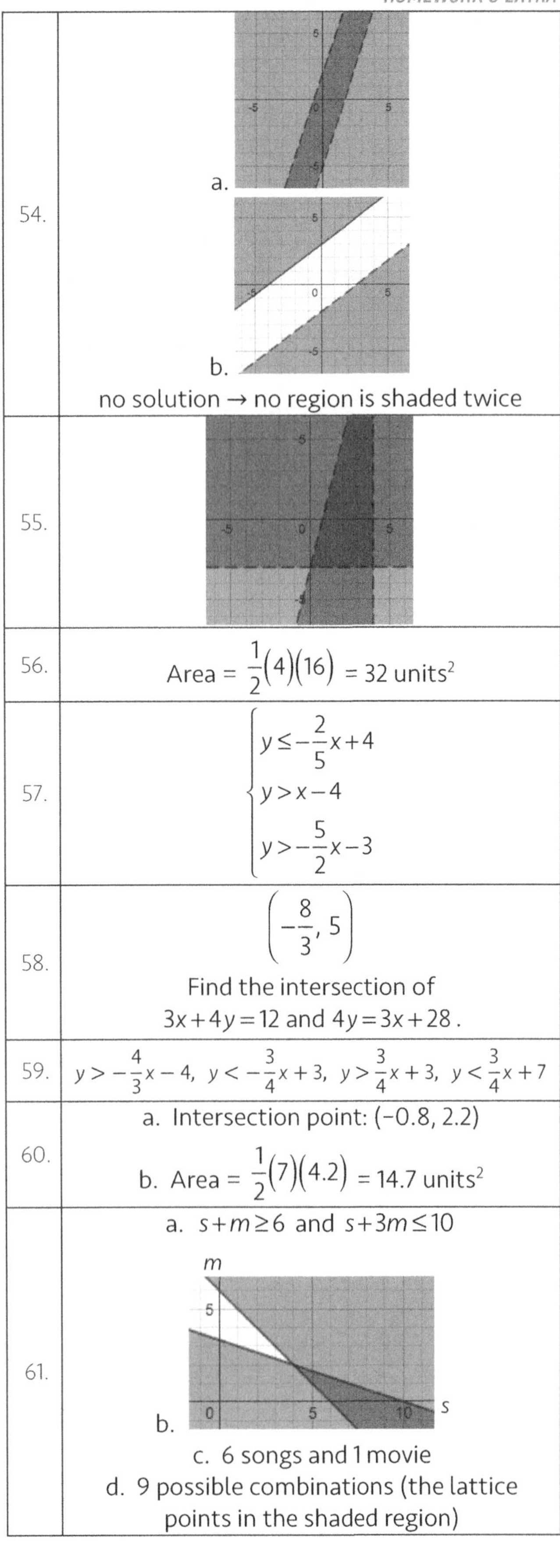

54.	a. b. no solution → no region is shaded twice
55.	
56.	Area = $\frac{1}{2}(4)(16)$ = 32 units^2
57.	$\begin{cases} y \le -\frac{2}{5}x+4 \\ y > x-4 \\ y > -\frac{5}{2}x-3 \end{cases}$
58.	$\left(-\frac{8}{3}, 5\right)$ Find the intersection of $3x+4y=12$ and $4y=3x+28$.
59.	$y > -\frac{4}{3}x - 4,\ y < -\frac{3}{4}x + 3,\ y > \frac{3}{4}x + 3,\ y < \frac{3}{4}x + 7$
60.	a. Intersection point: (−0.8, 2.2) b. Area = $\frac{1}{2}(7)(4.2)$ = 14.7 units^2
61.	a. $s+m \ge 6$ and $s+3m \le 10$ b. c. 6 songs and 1 movie d. 9 possible combinations (the lattice points in the shaded region)

62.	$y \le 3,\ y \ge -x-2,\ y \ge x-4$
63.	a. $x^2-22x+121$ b. $49y^2-42y+9$ c. t^4+2t^2+1
64.	$y^2+4y+4 \to$ If $y=x-2$, then $x=y+2$
65.	$-\frac{1}{2}(-6)^2-(-6)+10 \to -18+6+10 \to -2$
66.	x = 0, 2, or −2 $2x(x^2-4)=0 \to 2x(x+2)(x-2)=0$
67.	(−1, −3) and (5, 3)
68.	2 solutions (the circle, parabola and line all intersect at 2 points)
69.	Make the equations equal each other. First, solve for x: $-x^2-x+9=-x^2-3x+5$ Second, plug the x-value into one of the original equations to solve for y.
70.	$-x^2-x+9=-x^2-3x+5 \to x=-2$ $y=-(-2)^2-(-2)+9 \to y=-4+2+9 \to y=7$ The intersection point is $(-2,7)$.
71.	$x^2-2x+3=2x^2-7x-3 \to x^2-5x-6=0$ $\to (x-6)(x+1)=0 \to x=6$ or $x=-1$ $x=6 \to y=27 \to (6,27)$ $x=-1 \to y=6 \to (-1,6)$ Two intersection points: $(6,27)$ and $(-1,6)$.
72.	Solve for x: $x^2+3x=\frac{3}{2}x+1 \to 2x^2+3x-2=0$ (−2, −2) and (0.5, 1.75)
73.	(1,3), (0.5,3.5)
74.	(0,3), (3,6)
75.	(−8, 6) and (8, −6)
76.	24
77.	a. b.
78.	a. b. not possible

79.	$(-4, 3), \left(\frac{24}{5}, -\frac{7}{5}\right)$
80.	No solution.
81.	The equations represent two concentric circles. They do not intersect.
82.	$(8, 6), \left(-\frac{32}{13}, -\frac{126}{13}\right)$
83.	a. $R(15)=15{,}000 \rightarrow$ The company makes \$15,000 in revenue if it sells 1,500 cups. b. $C(23)=25{,}000 \rightarrow$ It costs the company \$25,000 to produce 2,300 cups.
84.	The company makes a profit if it sells between 500 and 2,000 cups. Between those n-values, the revenue line is higher than the costs curve.
85.	a. $x = 4 + 7y + 3z$ b. $y = z - 6x + 11$ c. $z = 5x + 5y - 1$
86.	a. $7y + z = -16$ b. $2x - 2z = -3$
87.	Many results are possible. One is shown. a. $3y - z = 9$ b. $-x + 26z = 17$ c. $17x + 7y = -15$
88.	(1, −2, 3)
89.	a. a plane, or a sheet of paper with edges that extend infinitely b. the common intersection point of 3 sheets of paper or 3 planes
90.	(−3, 2, 4)
91.	(1, −2, 3)
92.	(4, −3, 2)
93.	(0.5, 3, 7)
94.	a. 1 b. (1,1)
95.	Hint: Try the elimination method.
96.	Equation 1: $G + B + Y = 19{,}200$ Equation 2: $3G + 2B + 3.5Y = 56{,}400$ Equation 3: $G = 3B$ a. 4,800 b. Approx. 57.4% (\$32,400 out of \$56,400)
97.	$3=9A-3B+C$ $6=4A-2B+C$ $7=A-B+C$

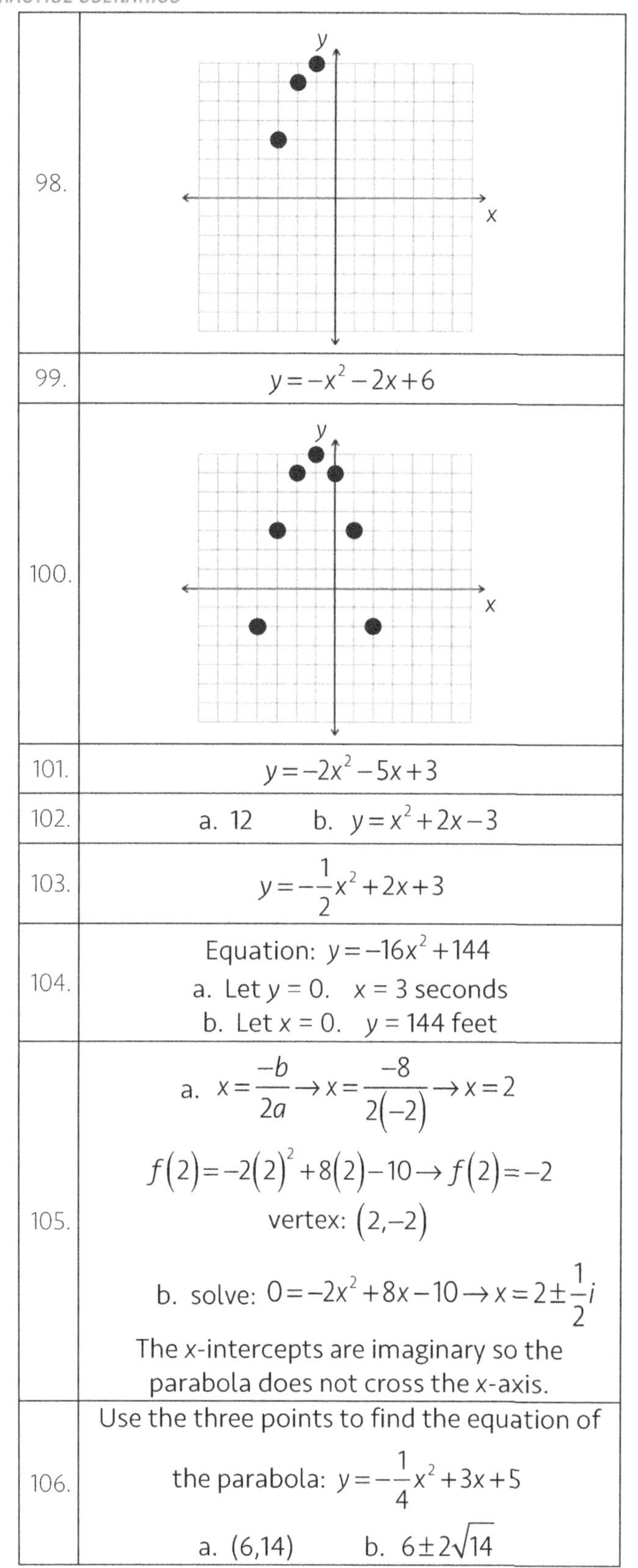

98.	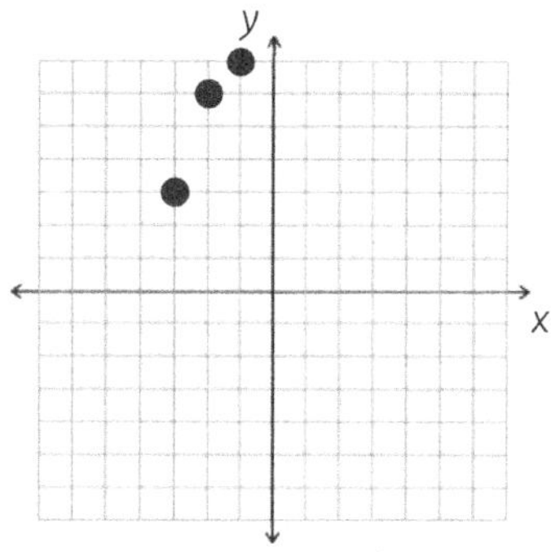
99.	$y=-x^2-2x+6$
100.	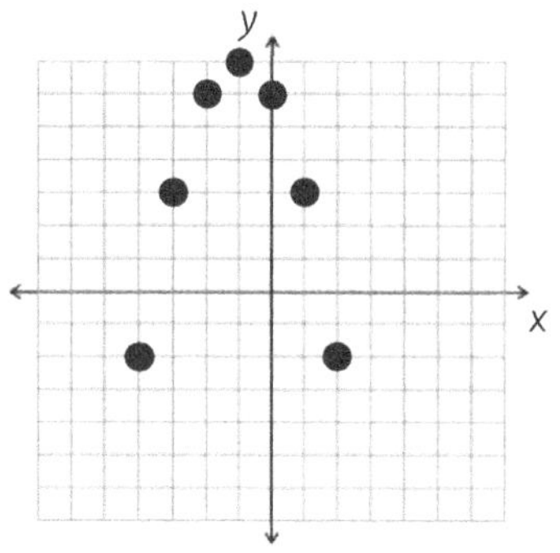
101.	$y=-2x^2-5x+3$
102.	a. 12 b. $y=x^2+2x-3$
103.	$y=-\frac{1}{2}x^2+2x+3$
104.	Equation: $y=-16x^2+144$ a. Let $y = 0$. $x = 3$ seconds b. Let $x = 0$. $y = 144$ feet
105.	a. $x=\frac{-b}{2a} \rightarrow x=\frac{-8}{2(-2)} \rightarrow x=2$ $f(2)=-2(2)^2+8(2)-10 \rightarrow f(2)=-2$ vertex: $(2,-2)$ b. solve: $0=-2x^2+8x-10 \rightarrow x=2\pm\frac{1}{2}i$ The x-intercepts are imaginary so the parabola does not cross the x-axis.
106.	Use the three points to find the equation of the parabola: $y=-\frac{1}{4}x^2+3x+5$ a. (6,14) b. $6\pm2\sqrt{14}$

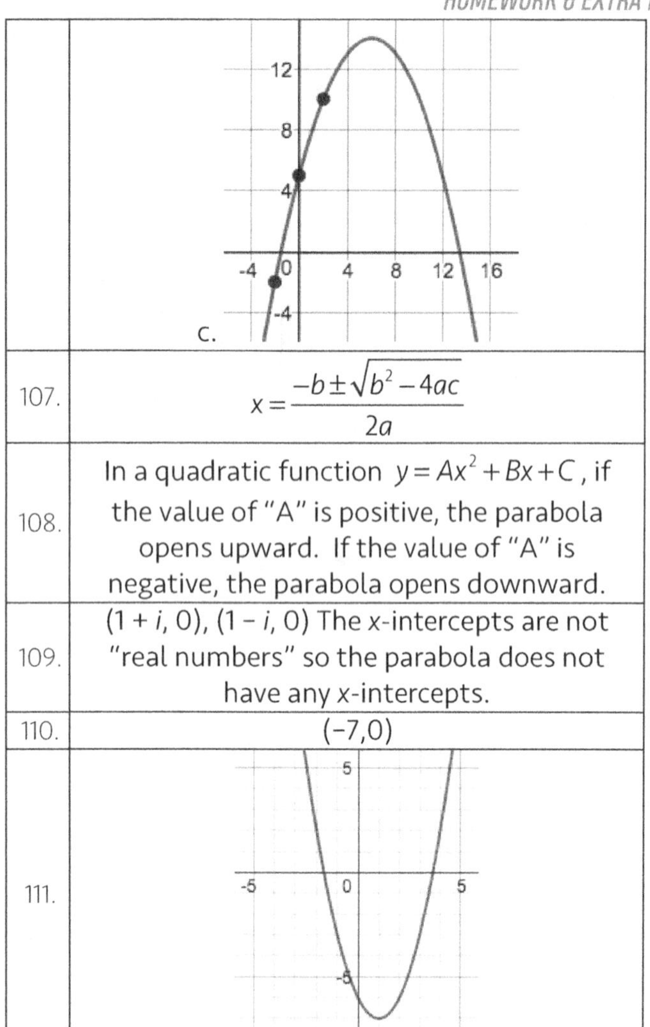

	c.
107.	$x = \frac{-b \pm \sqrt{b^2 - 4ac}}{2a}$
108.	In a quadratic function $y = Ax^2 + Bx + C$, if the value of "A" is positive, the parabola opens upward. If the value of "A" is negative, the parabola opens downward.
109.	(1 + i, 0), (1 − i, 0) The x-intercepts are not "real numbers" so the parabola does not have any x-intercepts.
110.	(−7,0)
111.	

112.	Base = 12 units Height = 6 units Area = 36 units^2
113.	2.4 gallons of vinegar solve: .2(4) + v = .5(4 + v)
114.	32 mins → 1920 seconds → 80 total dips → 6.4 gallons → .08 gallon per dip
115.	Derek's paycheck: P = 1000 + 0.10x Sonia's paycheck: P = 1300 + 0.05x Solve: 1000 + 0.10x = 1300 + 0.05x x = \$6,000 in sales Paycheck amount is \$1,600
116.	a. 1.2% b. 3.3%
117.	Equation 1: a + b =8,900 Equation 2: 0.05a + 0.012b =350 a = \$6,400 and b = \$2,500
118.	This point is the parabola's vertex. Its x-value is 3, halfway between the x-intercepts.
119.	m = 3 or −3

Made in the USA
Monee, IL
18 July 2024

62031058R00063